AF615394

COMPUTER SIMULATION OF MICROSTRUCTURAL EVOLUTION

Proceedings of a symposium sponsored by the American Society for Metals Materials Science Division Computer Simulation Technical Activity, held at the Fall Meeting of The Metallurgical Society in Toronto, Canada, October 13-17, 1985.

Edited By

David J. Srolovitz
Los Alamos National Laboratory
Los Alamos, New Mexico

A Publication of THE METALLURGICAL SOCIETY AIME The Metallurgical Society, Inc.

Library of Congress Cataloging-in-Publication Data

Computer simulation of microstructural evolution.

Includes index.
1. Metallography--Mathematical models--Congresses.
2. Metallography--Data processing--Congresses.
I. Srolovitz, David J. **II.** American Society for
Metals. Materials Science Division. Computer Simulation
Technical Activity. **III.** Metallurgical Society of
AIME. Fall Meeting (1985 : Toronto, Ont.)
TN689.2.C66 1986 669'.95'0724 86-8687
ISBN 0-87339-018-0

A Publication of The Metallurgical Society, Inc.
420 Commonwealth Drive
Warrendale, Pennsylvania 15086
(412) 776-9000

Printed in the United States of America.
Library of Congress Catalogue Number 86-8687
ISBN NUMBER 0-87339-018-0

Preface

While computer simulation is fairly widespread in materials science, most of the activity has concentrated on the electronic or atomic levels. Electronic structure calculations are either performed for small clusters of atoms (< 100) or for perfect crystals where the number of atoms employed in the calculation is equal to the number of atoms in a unit cell. Atomistic calculations typically employ between $10^2 - 10^4$ atoms. While these numbers are increasing with improvements in available techniques and computer technology, we are still far from the day when such simulations can be performed on a scale sufficiently large to include much more than a single, isolated microstructural feature (e.g. grain boundaries, dislocations, etc.). For example, in order to simulate one cubic micron of a metal (still a rather fine microstructural scale) one must simulate of order 10^{11} atoms. This is beyond the current state of the art by approximately seven orders of magnitude. Because of this disparity between the size scales on which significant microstructural complexity exists and the inherent size limitations of atomistic simulations, microstructural simulation must operate on a nonatomistic basis. These limitations of the atomistic approaches are responsible, in part, for the development of the many new approaches to microstructural simulations described herein.

The majority of the computational techniques described in these papers simulate the temporal evolution of microstructures by time integration of equations describing the microstructural features of interest. In a number of cases these equations are simply empirical relations which describe how a given microsturcture evolves as a function of such parameters as quench rate, temperature, etc. These types of models are often coupled to local descriptions of the evolution of these parameters; for example, thermal diffusion accompanying solidification. Such local descriptions may be applied directly to the evolution of the microstructural features by employing the equations of motion for small segments of the microstructural feature of interest; for example, grain boundaries, solidification fronts, etc. Depending on the information required to simulate the evolution of the microstructure, such varied techniques as finite elements, finite differences, front tracking, and boundary integral methods have been applied. A fundamentally different approach to the simulation of microstructural evolution is based on a local description of the material and simple evolutionary rules, instead of the usual differential and integral equations. These rules may simply be algorithmic, as in the case of cellular automata, or stochastic, as in the case of Monte Carlo simulations. While inherently discrete, such approaches may approximate continuum systems and are generally more computationally efficient than the integration approach.

The papers in this volume were presented at an international symposium held during the 1985 Fall Meeting of The Metallurgical Society of AIME in Toronto and sponsored by the Computer Simulation Activity of the Materials Science Division of the ASM. The purpose of this symposium was to bring together researchers who work in the rapidly growing area of computer simulation of microstructures. The timing for this symposium was particularly opportune due to the recent development of new techniques in the analytic, numerical and simulational aspects of microstructural evolution. In order to increase the utility of this volume, many of the authors have explicitly and pedagogically described the techniques which they employed.

David J. Srolovitz
Los Alamos National Laboratory
Los Alamos, New Mexico

January, 1986

Table of Contents

COMPUTER SIMULATION OF BIDIMENSIONAL GRAIN BOUNDARY MIGRATION (I):

THE ALGORITHM

E. A. Ceppi and O. B. Nasello

Grupo de Física de la Atmósfera
Facultad de Matemática, Astronomía y Física
Universidad Nacional de Córdoba
Laprida 854, 5000 Córdoba - Argentina

Abstract

A computer algorithm which simulates grain boundary migration is developed on the basis of geometrical arguments. The simulation model takes into account the drag effect of dispersed second phase particles (Zener drag) and the difference of bulk free energies between crystals. A discussion of the improvements of the algorithm in order to consider non-isotropic grain boundary energies is made. Simulation results are given and discussed thereby checking the model's performance.

Introduction

Grain boundary migration occurs when a grain boundary is subjected to a driving force sufficient to cause the motion of the boundary. This migration process causes a reduction in the free energy of the system.

In a fully recrystallized polycrystalline material, the energy stored in grain boundaries provides the driving force for migration. This process, known as grain growth, is then characterized by the reduction of the grain boundary surface and, consequently, by an increment of the mean grain size. The kinetics of grain growth is a consequence of the migration of individual grain boundaries. Then, the solution of the equation of motion of each boundary in the sample can give us the necessary connection between the individual and collective processes. This connection can be found as follow.

Consider a sample composed of N grains and let Γ_{ij} be the grain boundary area connecting grains "i" and "j" (i,j = 1,.,N). If V_i is the volume of crystal "i", then the growth rate of V_i due to the migration of Γ_{ij} is given by:

$$\left.\frac{dV_i}{dt}\right|_j = \int_{\Gamma_{i,j}} \vec{V_L} \cdot \hat{n}\, dS \qquad (1)$$

where $\vec{V_L}$ is the local grain boundary velocity and $\hat{n}$ is the normal to Γ_{ij}, directed outward grain "i".

Equation (1) can be written as:

$$\left.\frac{dV_i}{dt}\right|_j = v_{i,j}\, A_{i,j} \qquad (2)$$

where $A_{i,j}$ is the area of Γ_{ij} and $v_{i,j}$ is the mean grain boundary velocity defined by:

$$v_{i,j} = (1/A_{i,j}) \int_{\Gamma_{ij}} \vec{V_L} \cdot \hat{n}\, dS \qquad (3)$$

Therefore, the equations of evolution of the polycrystalline system are:

$$\frac{dV_i}{dt} = \sum_{j=1}^{N} v_{i,j}\, A_{i,j} \qquad i = 1,\ldots, N \qquad (4)$$

Simultaneous integration of the system of equations (4) were made by different authors [1,7] using some assumptions about the magnitude of $v_{i,j}$ and $A_{i,j}$. In all cases, a capillary driving force was used and the shape of the grains was supposed nearly spherical. In some papers [1], [7], the influence of dispersed second phase particles (Zener drag) was also studied.

The simplifying assumptions made in order to solve (4) make if difficult to apply the results obtained with the models to the interpretation of experimental results from polycrystalline samples with irregular grains. For example, this interpretation problem arises in some studies of grain growth in natural ice samples [8] and [9].

Another approach to the problem is provided by computer simulation models of grain boundary migration [10,12]. These models make it possible to obtain the sample evolution without solving directly system (4), since they are able to simulate each boundary movement in the sample. Therefore, the simplifying assumptions concerning the shape of the crystals are not required.

The model proposed by Srolovitz et al. [10,11] is based on statistical arguments and makes use of a Monte Carlo technique to obtain the system evolution. On the other hand, the model proposed by us in a previous paper [12] makes use of geometical arguments to solve the equation of motion of each grain boundary.

The present work will give a generalization of the latter model which can be applied to bidimensional samples containing recrystallized grains and dispersed second phase particles. Part I will be devoted to defining the algorithm and to show how it works in some bicrystalline examples. In part II a study is made on the effect of Zener drag on grain growth and then the evolution of a real sample is compared with that given by the simulations model.

Algorithm

Fig. 1 shows a bidimensional sample composed by two crystals A and B whose interface is designated by Γ_{AB} . The following functions of the coordinate x are defined:

$$f(x) = \begin{cases} 1 \text{ if } x \text{ belongs to A} \\ -1 \text{ if } x \text{ belongs to B} \\ 0 \text{ if } x \text{ does not belong to A and B} \end{cases}$$

$$F(x) = \int_{C(x,a)} f(x)\, d^2x \qquad (5)$$

where the integration area $C(x,a)$ is a circle of radius "a" centered at x.

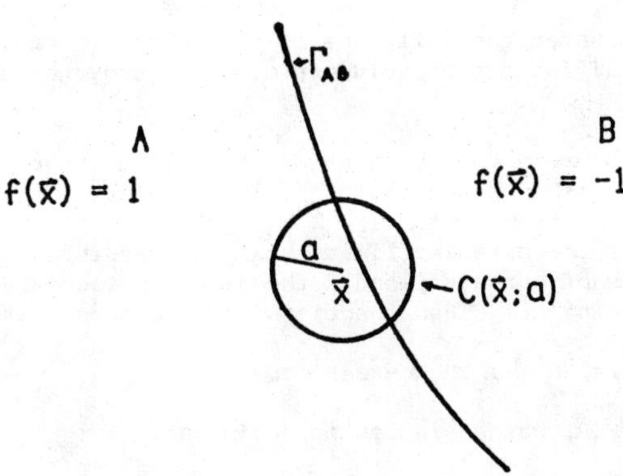

Figure 1 - Grain boundary between crystals A and B and integration area C(x,a).

If U_{AB} is a function of x for which $|U_{AB}| \ll \pi a^2$; the curve Γ_{AB} implicitly defined by:

$$F(x) = U_{AB}(x) \qquad (6)$$

represents the position of the boundary Γ_{AB} after a time interval $\Delta t = a^2/(6\ M\ \sigma_{gb})$, according to the equation of motion:

$$V_L = \frac{M\ \sigma_{gb}}{\rho} - \frac{3}{2}\ \frac{M\ \sigma_{gb}}{a^3}\ U_{AB}(x) \tag{7}$$

Equation (7) is found to first order in (a/ρ).

The appropiate choice of the function U_{AB} gives different cases of interest. For example:

a) $U_{AB} = 0$

$$V_L = -\frac{M\ \sigma_{gb}}{\rho} \tag{8}$$

the grain boundary is only driven by capillary forces.

b) $U_{AB} = -\ (2a^3/3)\ f(x)\ Z$

$$V_L = -\frac{M\ \sigma_{gb}}{\rho} \pm M\ \sigma_{gb}\ Z \tag{9}$$

the grain boundary is capillary driven but a dispersed second phase particles drag Z (Zener drag) is included; the (±) sign indicates that the drag is always opposed to the movement.

c) $U_{AB} = -\ (2a^3/3\ \sigma_{gb})\ (g_B - g_A)$

g_A and g_B are the volumetric free energies of regions A and B, respectivelly.

In this case:

$$V_L = -\frac{M\ \sigma_{gb}}{\rho} + M\ (g_B - g_A) \tag{10}$$

the boundary moves under the influence of two forces: a capillary one and that given by the difference of volumetric free energy between crystals A and B.

The proposition expressed by equations 5 to 7 can be proved without loss of generality if it is supposed that Γ_{AB} is circular.

Figure 2 shows the circular limit Γ_{AB} and the integration area $C(x,a)$. Let R be the radius of curvature and R the distance from the center of curvature x_o to the point x. Then function $F(x)$ can be written as:

$$F(x) = F(R') = 2a^2\ \beta + 2R^2\ \phi - 2aR'\ \cos\ \phi \tag{11}$$

The function $F(R')$ has an inflection point at:

$$R' = R_o = +\ (R^2 - a^2)^{1/2}$$

then:

$$\left.\frac{\partial^2\ F}{\partial^2\ R'}\right|_{R_o} = 0 \tag{12}$$

Therefore, $F(R')$ can be aproximated by:

$$F(R') = F(R_o) + \left.\frac{\partial\ F}{\partial R'}\right|_{R_o} (R' - R_o) \tag{13}$$

in the neighborgood of R_o, neglecting third order terms in $(R' - R_o)$.

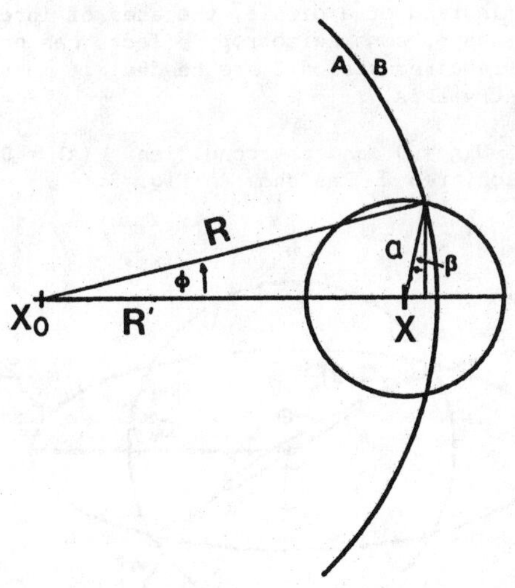

Figure 2 - Circular grain boundary between crystals A and B.

Expanding R_o to first order in (a/R), then the expression (13) may be written as:

$$F(R') = - 4a \left[(a^2/6R) + (R' - R) \right] \quad (14)$$

Thus, condition (6) is equivalent to:

$$\delta R = (R' - R) = - \frac{a^2}{6R} - \frac{U_{AB}}{4\,a} \quad (15)$$

to first order in (a/R). If δt is defined as:

$$\delta t = a^2/(6M\,\sigma_{gb}) \quad (16)$$

then equation (12) can be expressed as:

$$\frac{\delta R}{\delta t} = - \frac{M\,\sigma_{gb}}{R} - \frac{3}{2}\,\frac{M\,\sigma_{gb}}{a^3}\,U_{AB} \quad (17)$$

which is equivalent to (3), taking into account that in this case $R = \rho$.

It must be noted here that the product $M\sigma_{gb}$ is supposed to be isotropic, independent of boundary inclination and of crystal orientation. However, as will be seen, anisotropy can be included in this model.

Algorithm extensions

In this section, the proposed algorithm will be extended in order to take into account the anisotropy in grain boundary migration and to apply the algorithm to polycrystalline samples.

Anisotropy in $M\sigma_{gb}$

Function F(x) is defined in (1) by an integration over a circular area of function f(x). If instead of a circle, the area of integration Σ is a zone with a different shape, some anisotropy effects can be taken into account. However, some requirements on Σ are needed; it must have at x at least a two fold symmetry axis.

Let suppose that $U_{AB} = 0$ and the condition $F(x) = 0$ is fulfilled at x, for an integration area Σ, as show in Fig. 3.

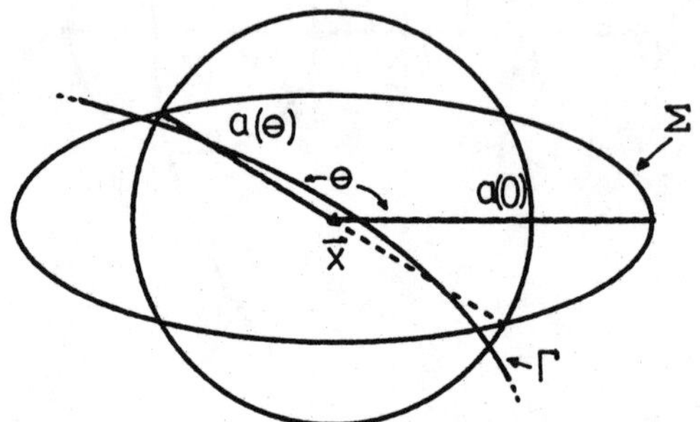

Figure 3 - Condition $F(x) = 0$ fulfilled at a non circular integration area Σ and at a circle of $a(\theta)$ radius.

It can be proved that there will always be a straight line of length 2 $a(\theta)$ that allows for the same condition to be fulfilled by a circular integration area of radius $a(\theta)$. Therefore, as showed in (15), the distance of x from Γ_{AB} is:

$$\delta R = - a(\theta)^2/(6\rho) \tag{18}$$

If δt is defined as:

$$\delta t = a(0)^2/(6 M\sigma_{gb}(0)) \tag{19}$$

then, in this case, equation (17) may be written:

$$\frac{\delta R}{\delta t} = - \frac{M \sigma_{gb}(\theta)}{\rho} \tag{20}$$

with

$$M\sigma_{bg}(\theta) = M\sigma_{gb}(0) \, [a(\theta)/a(0)]^2 \tag{21}$$

Thus, if M is considered isotropic, the Wulff plot of $\sigma_{gb}(\theta)$ is related with the polar plot of $a^2(\theta)$.

Polycrystals

The algorithm could be generalized to be applied to a polycrystalline sample composed of N crystals in the following way. Funtions $f_n(x)$ and $F_n(x)$ are defined:

$$f_n(x) = \begin{cases} 1 \text{ if } x \text{ belongs to crystal "n"} \\ 0 \text{ if } x \text{ does not belong to crystal "n"} \end{cases} \quad (n = 1,\ldots, N)$$

$$F_n(x) = \int_{C(x,a)} f_n(x)\, d^2x \qquad (22)$$

Furthermore, the functions $U_n(x)$ are defined with the condition $|U_n(x)| \ll \pi a^2$. Thus, if Γ is the surface that represent all the grain boundaries in the sample, the configuration after a time interval $\delta t = a^2/6 M \sigma_{gb}$ is represented by the surface Γ':

$$\Gamma' = \{x / F_n(x) - F_m(x) = U_n(x) - U_m(x), \quad m, n = 1,\ldots, N\}$$

Γ' corresponds to the set of all the centers of $C(x,a)$ in which the relation is fulfilled. In particular when $U_n(x) = 0$ Γ' is the set of all centers of $C(x,a)$ for which the intersection areas between grains and circle are equal. Others choices of U_n gives the different cases of interest.

Examples

Anisotropy in grain boundary migration

Fig. 4 shows the evolution of an initially circular crystal embedded in a larger one of different orientation. The shape of the integration area Σ was selected in order to simulate the orientation dependence of σ_{gb} in a rectangular lattice with a relation $a/b = 1,36$ between its parameters (see Fig. 5).

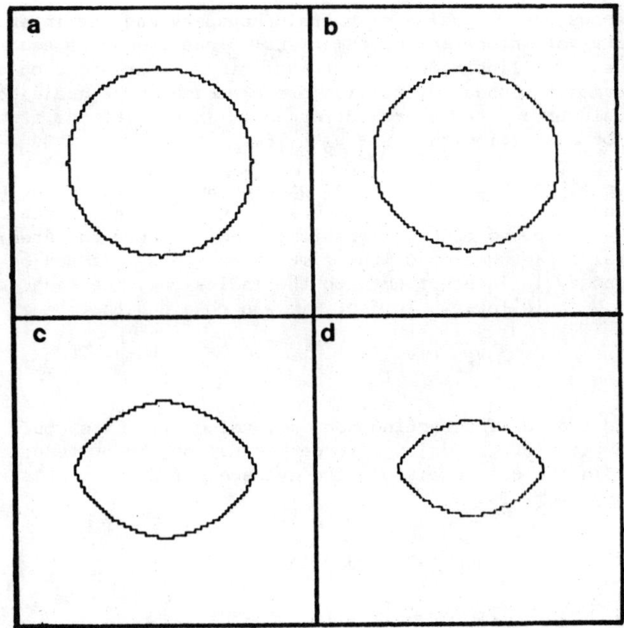

Figure 4 - Crystal evolution with anisotropic $M\,\sigma_{gb}$:
a) p = 0 ; b) p = 5 ; c) p = 25 ; d) p = 45.

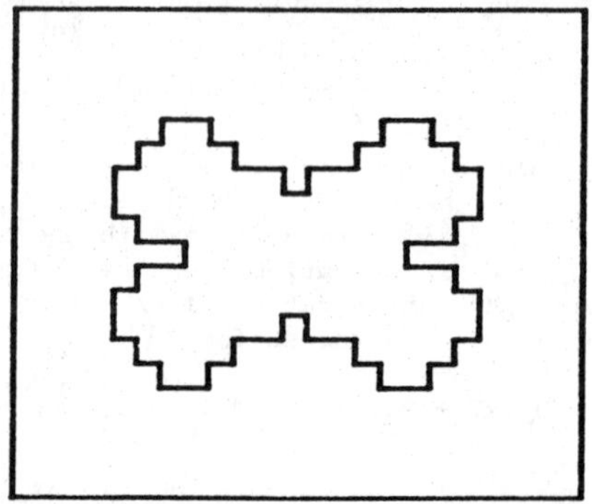

Figure 5 - Integration area Σ used to found the crystal evolution shown in Fig. 4 (maximun length 15 lattice points. maximun width 11 lattice points).

In Fig. 4 it can be seen that the circle shrinks, changing it shape to an oblate one. This is the result of the higher values of $M\,\sigma_{gb}$ for lattice planes with lower densities of lattice points. It is interesting to note that some faceting is observed at the characteristic Wulff planes of the $M\,\sigma_{gb}$ used.

Bypassing of inclusions

Fig. 6 shows the migration of a grain boundary and its interaction with a periodic array of incoherent circular inclusions of radius equal to seven lattice points. The linear density of inclusions is one to each 64 lattice points and periodic boundary condition are employed. The grain boundary is driven by a volumetric free energy diference. In eq. (6) the function U_{AB} used is that of case (c) with $g_A - g_B = 16$.

In Fig. 6 it is observed that the pinning action of the inclusions makes the grain boundaries curved. Then, there is an increment of the capillary forces which are opposed to the migration. Fig. 7 shows the area of the growing crystal as a function of the simulation steps. It can be seen that, while the boundary is interacting with the inclusions, the mean grain boundary velocity V_{LI} is lower. In fact, it was obtained that:

$$V_{LI} = V_{LO}/3$$

where V_{LO} is the non interacting boundary velocity. This could be interpred as a reduction of the driving force because of the presence of inclusions. Then, in the case analysed, the average effective back stress acting on the boundary Δg_b is:

$$\Delta g_b = -\,2\,\Delta g/3$$

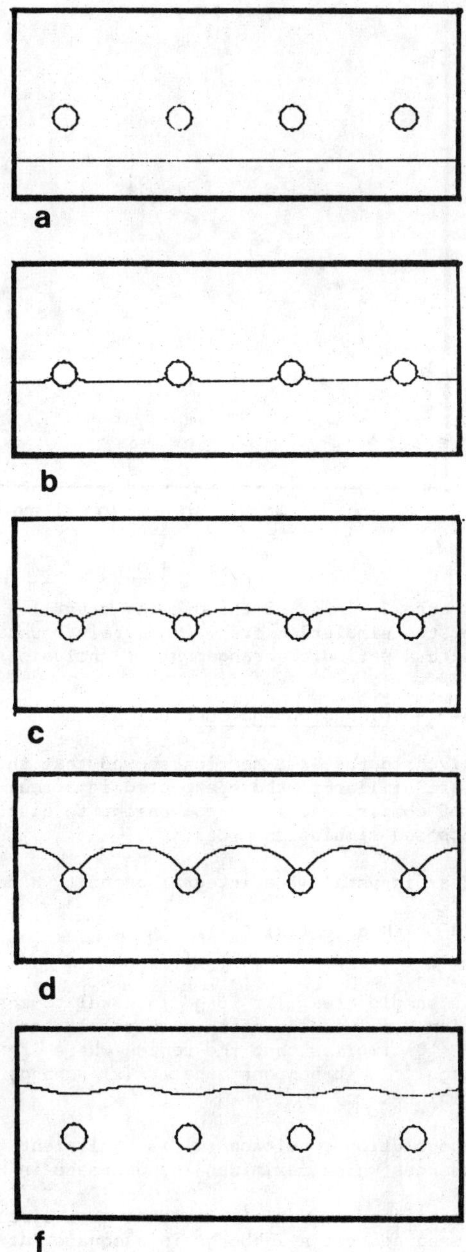

Figure 6 - Grain boundary interaction with a periodic arrangement of circular inclusions: a) to f) simulation step 0 to 80 step 20.

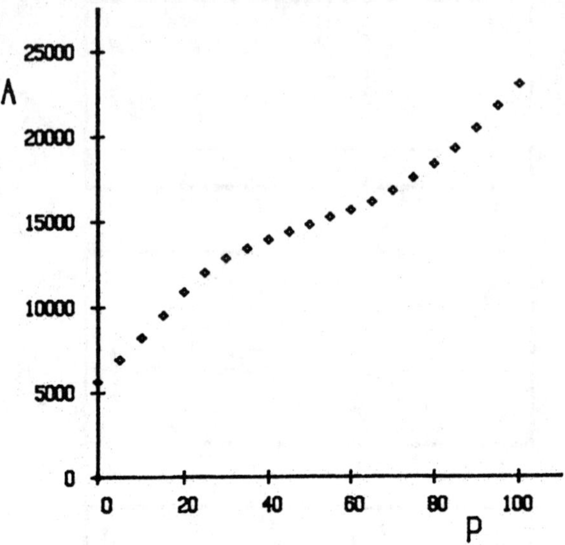

Figure 7 - Area of the growing crystal as a function of the simulation step. The grain boundary interact with a periodic arrangement of inclusions.

Discusion and conclusions

The examples given in the last section showed that the results obtain with the algorithm are similar to those expected in a real experiment. However, for the sake of comparison, it is convenient to discuss some aditional features of the propposed simulation model.

In the case of a bicrystal, the integral quantity H can be defined by:

$$H = \int_S [\pi a^2 - f(x)\ F(x)]\ d^2x \qquad (23)$$

where S is the whole sample area. If U_{AB} is equal to zero and isotropy is supposed, the function H is reduced after each simulation step. In fact, the condition $F(x) = 0$ implies that the region where $f(x)\ F(x) < 0$ must change its orientation. If it happens that $f(x)$ change its sign, the integrand in (23) diminishes.

Therefore, the evolution condition (2) is equivalent to finding all of the orientational changes which maximizes the decrease in H.

When the integration area $C(x,a)$ is taken as a circle of radius equal to the distance between nearest neighborgs in a square lattice, the quantity H is equivalent to the "Ising" like Hamiltonian used in the model of Srolovitz et al. [10,11]. In this case, our model could be compared with that of the quoted authors in the limit of $T = 0$. However, no changes are detected at this temperature by these authors and no curvature can be measured by such an small integration area in the present model.

On the other hand, it must be pointed out that our model is intended to detect the boundary curvature inside C(x,a) and to move the limit according to the grain boundary equation of motion in a deterministic way. No attemps were made in order to include fluctuations or any other statistical effects in the model.

Acknowledgments

The financial support of the Consejo Nacional de Investigaciones Científicas y Técnicas and the Consejo de Investigaciones Científicas y Tecnológicas de la Provincia de Córdoba is gratefully acknowledged.

The authors are indebted to Dra. Laura Levi and to Dr. Jorge M. Caranti for their helpful comments.

References

1. P. Feltham, Acta Metallurgica, 5 (1957) 97.

2. M. Hillert, "On the Theory of Normal and Abnormal Grain Growth", Acta Metallurgica, 13 (1965) 227 - 238.

3. N.P. Lovat "On the Theory of Normal and Abnormal Grain Growth", Acta Metallurgica, 22 (1974) 721 - 724.

4. V.Y. Novikov, "Computer Simulation of Normal Grain Growth", Acta Metallurgica, 26 (1978) 1739 - 1744.

5. V.Y. Novikov, "On Computer Simulation of Texture Development in Grain Growth", Acta Metallurgica, 23 (1979) 1461 - 1466.

6. O. Hunderi and N. Ryum, "Computer Simulation of Grain Growth", Acta Metallurgica, 27 (1979) 161 - 165.

7. O. Hunderi and N. Ryum, "Computer Simulation of Stagnation in Grain Growth", Acta Metallurgica, 19 (1981) 1737 - 1745.

8. F. Prodi and L. Levi, "Aging of Accreted Ice", Journal of the Atmospheric Sciences, 37 (1980) 1376 - 1384.

9. L. Levi and E.A. Ceppi, "Grain Growth in Ice". Il Nuovo Cimento, 5C (1982) 445 - 461.

10. D.J. Srolovitz, M.P. Anderson, G.S. Grest and P.S. Sanhi, "Grain Growth in two Dimensions", Scripta Metallurgica, 17 (1983) 241 - 246.

11. M.P. Anderson, D.J. Srolovitz, G.S. Grest and P.S. Sanhi, "Computer Simulation of Grain Growth" - I. Kinetics, Acta Metallurgica, 32 (1984) 783 - 791.

12. E.A. Ceppi and O.B. Nasello, "Computer Simulation of Bidimensional Grain Growth", Scripta Metallurgica, 18 (1984) 1221 - 1225.

COMPUTER SIMULATION OF BIDIMENSIONAL GRAIN BOUNDARY MIGRATION (II):

AN APPLICATION TO GRAIN GROWTH

O. B. Nasello and E. A. Ceppi

Grupo de Física de la Atmósfera
Facultad de Matemática, Astronomía y Física
Universidad Nacional de Córdoba
Laprida 854, 5000 Córdoba - Argentina

Abstract

The algorithm developed to simulate grain boundary migration is applied to bidimensional polycrystalline samples. The kinetics of grain growth is analyzed and the effect of a dispersion of second phase particles is especially discussed. The results are compared with theoretical predictions and with experimental data from ice polycrystals.

Introduction

The study of grain growth in ice is particularly interesting in Cloud Physics and Glaciology. As it can be seen from some papers [1-3], the polycrystalline samples of interest are generally composed of crystals with complex shapes. Moreover, air bubbles are present and distributed in a non-uniform pattern. The quoted characteristics make it difficult to apply the grain grouwth models developed up to now [4-7], since they are restricted to polyedral shaped crystals. On the other hand, the way to compare the kinetic parameters from samples with different bubble concentration is not clearly stated.

These problems make it necesary to develop a different kind of approach to link the single grain migration properties with the average kinetics of grain growth.

In a previous paper [8], we presented a model which simulates the grain boundary migration in bidimensional samples under capillary driving forces. An improvement of this model, in order to consider anisotropy and other driving forces was proposed in part I of the present work.

The modified model will be used here in two different ways:

a) to obtain some general results in case of grain growth with Zener drag,

b) to compare the temporal evolution of a real sample and the corresponding simulated microstructure, in order to find a value of the parameter $M\sigma_{gb}$.

The model characterization

The model developed in part I can be used to study the evolution of a polycrystalline sample in which the grain boundary velocity is given by:

$$V_L = -\frac{M\sigma_{gb}}{R} - \frac{3}{2}\frac{M\sigma_{gb}}{a^3}U_n(x) \ .$$

Normal grain growth with capillary driving forces is simulated taking $U_n(x) = 0$ ($n = 1,\dots,N$; N is the number of grains in the sample).

The influence of dispersed second phase particles on grain growth can be analysed with the present model in a continuous way by choosing

$$U_n(x) = \begin{cases} -\dfrac{2a^3}{3}Z & \text{with } Z > 0, \text{ if } x \text{ belongs to crystal } n. \\ 0 & \text{in any other case.} \end{cases}$$

Results

Grain growth with Zener drag

The temporal evolution of the mean grain size D of a computer generated sample is shown in Fig. 1. Different drag levels α ($\alpha = 2a^2 Z/3$) were used. In all cases, except for $\alpha = 0$, the grain growth process slows down until it is completely stopped. The limiting grain size is higher for smaller values of α, and the time required for the system to become pinned increases for decreasing α.

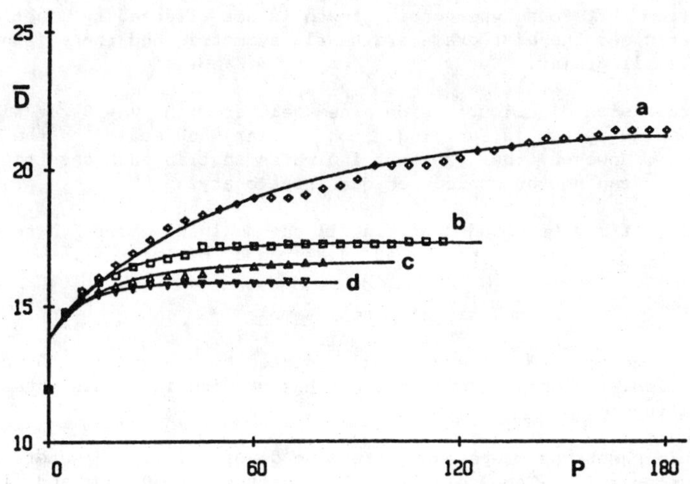

Figure 1 - Mean grain size $\bar{D}$ vs. simulation step p, for different values of the drag parameter α: a) $\alpha = 1$; b) $\alpha = 2$; c) $\alpha = 3$; d) $\alpha = 5$.

The solid lines in Fig. 1 represent the fitting of the data by the expression proposed by Burke (see [9]).

$$\frac{Kp}{D_s} = \frac{D_o - \bar{D}}{D_s} + \ell n \left[\frac{D_s - D_o}{D_s - \bar{D}}\right] \quad (1)$$

where D_s is the limiting mean grain size and D_o is the initial mean grain size.

When $\alpha = 0$, the grain growth kinetics is expressed, for long simulations ($p \geqslant 100$ of time steps), by the quadratic law:

$$\bar{D}^2 - \bar{D}_o^2 = k_2\ p \quad (2)$$

It is interesting to note that, in the cases analysed, the parameter K in eq. (1) is approximately related to k_2 by the expression:

$$K(D_s) = \frac{k_2}{2\ (1 - D_o/D_s)} \quad (3)$$

The equation (3) allows experimental results from samples of the same material to be compared directly, even when they have different Zener drag.

Fig. 2 shows the normalized grain size distribution obtained from simulations with different values of α , when the grain growth process has stagnated. These histograms resembles that found by Hundery and Ryum [7] and Srolovitz et al. [8] at stagnation, ie. the final distributions show no grains smaller than 0.5 $\bar{D}$, then it increases rapidly, reaches a maximun and falls slowly towards large grain sizes. The shape of the distribution function is different from that found when grain growth is not affected by Zener drag. In the later case, the histograms are nearly symmetric and there is no absence of small grains.

The grain size distributions obtained, either with $\alpha = 0$ or with $\alpha \neq 0$, when the system is evolving, look similar when scaled by the mean grain size $\bar{D}$. However, the χ^2 test indicates that in each case the grain growth process can not be considered quasistationary.

Finally, it may be noted that, in all the analysed cases, there was no evidence of abnormal grain growth, as suggested by Hillert [4].

Analysis of a real sample

In order to compare the evolution predicted by the model with that found in real samples, the grain growth process has studied in a thin sheet of polycrystalline ice.

The thin sheet was prepared by freezing double distilled water on a cold microscope slide. The ice sheet has a thickness of 0.2 mm and a initial mean grain size of 0.1 mm. No bubbles were observed in the sample as inspected under a transmision microscope.

The sample was immersed on a microscope cold stage thermostatized to -2°C and filled with silicone oil to prevent evaporation. The temporal evolution of the crystalline structure was obtained by photographing at regular intervals using a polarizing transmision microscope.

Fig. 3 shows the time evolution of the square of the mean grain size $\bar{D}$ and the fitting by the expression:

$$\bar{D}^2 - D_o^2 = k_2 \ t$$

The values obtained for the parameters D_o and k_2 are

$$D_o = 0{,}09 \text{ mm}$$
$$k_2 = 3{,}3 \times 10^{-6} \text{ mm}^2 \text{ seg}^{-1}$$

This behaviour suggest that, within the time interval of the experiment, there were negligible drag effects.

In order to provide the initial microstructure of the system for the simulation model, a region of the sample after 2 hours of annealing was digitized. The length scale was chosen to be 0,49 mm/(lattice site). The simulation was carried out using isotropic conditions on $M\,\sigma_{gb}$ and zero values for all $U_n(x)$ functions in eq. (1).

Careful comparison between simulated and real evolution indicates that for $\Delta p = 10$ steps and $\Delta t = 2$ hours the microstructures are quite similars. As observed in Fig. 4, in general there is a good agreement between simulated and real evolutions, but same discrepancies can be found.

In part I it was shown that $\Delta t = (a^2/6\,M\sigma_{gb})\ \Delta p$.

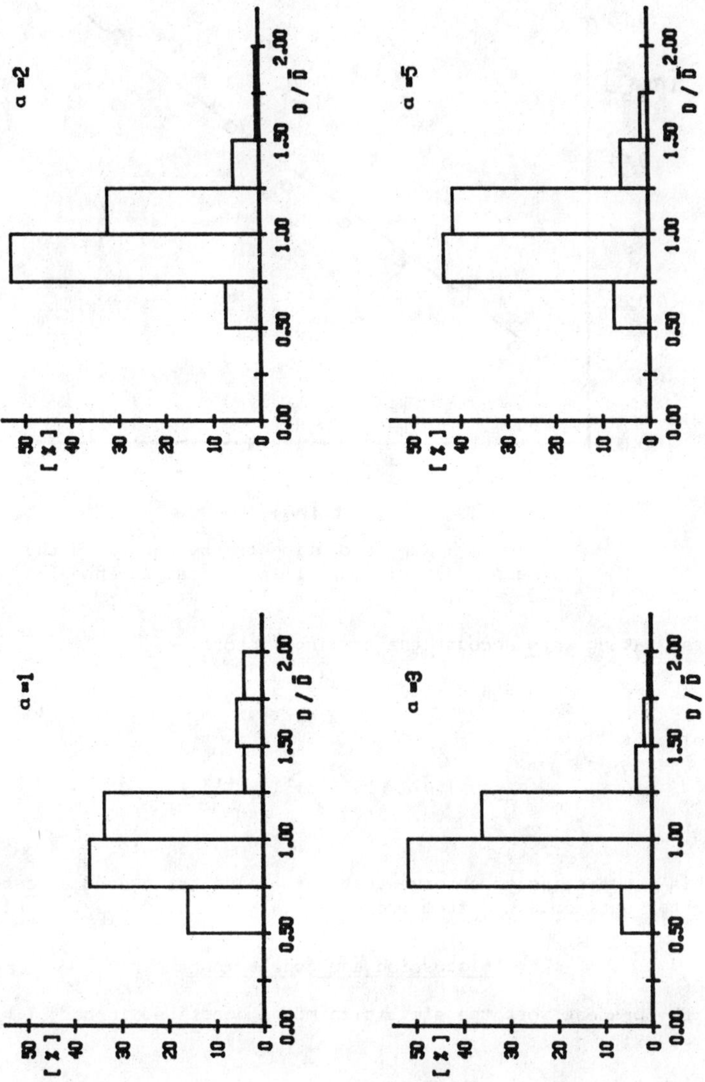

Figure 2 - Normalized grain size (D/D̄) histogram for different values of the drag parameter α , at stagnation.

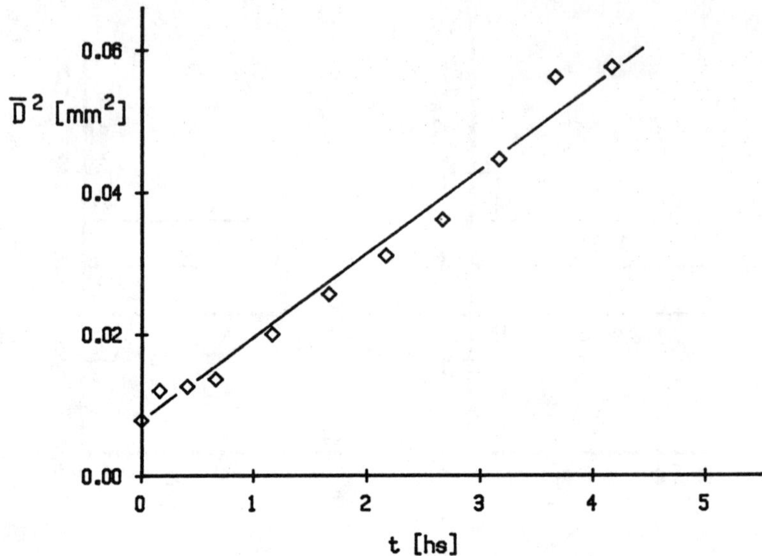

Fig. 3 - Time evolution of the square of the mean grain size $\overline{D}$ of an ice sample annealed at -2°C.

Therefore, taking into account the scaling factors,

$$a = 6{,}8 \times 10^{-3} \text{ cm}$$

and a value of

$$M\sigma_{gb} = 1{,}07 \times 10^{-8} \text{ cm}^2 \text{ seg}^{-1}$$

is found.

It is interesting to observe that this value is in good agreement with experimental data obtained from ice bicrystals annealed at -2°C [11].

Discussion and conclusions

In the present work the simulation model described in part I has been used to study:

1 - Grain growth with Zener drag.
2 - Crystal structure evolution of a real sample.

In the first case, the observed behavior is similar to that obtained in grain growth experiments with samples having a dispersed second phase [3]. In fact, the mean grain size reaches a higher limit value for a lower drag parameter α, which is proportional to the fraction f of the second phase.

The grain growth kinetics has been fitted satisfactorally by the Burke expression (eq. 1). Also, a relation was found between the kinetics parameters K and k_2, given in equations (1) and (2). The importance of this relation is that it makes possible to compare grain boundary mobilities obtained from samples of the same material but with different concentration

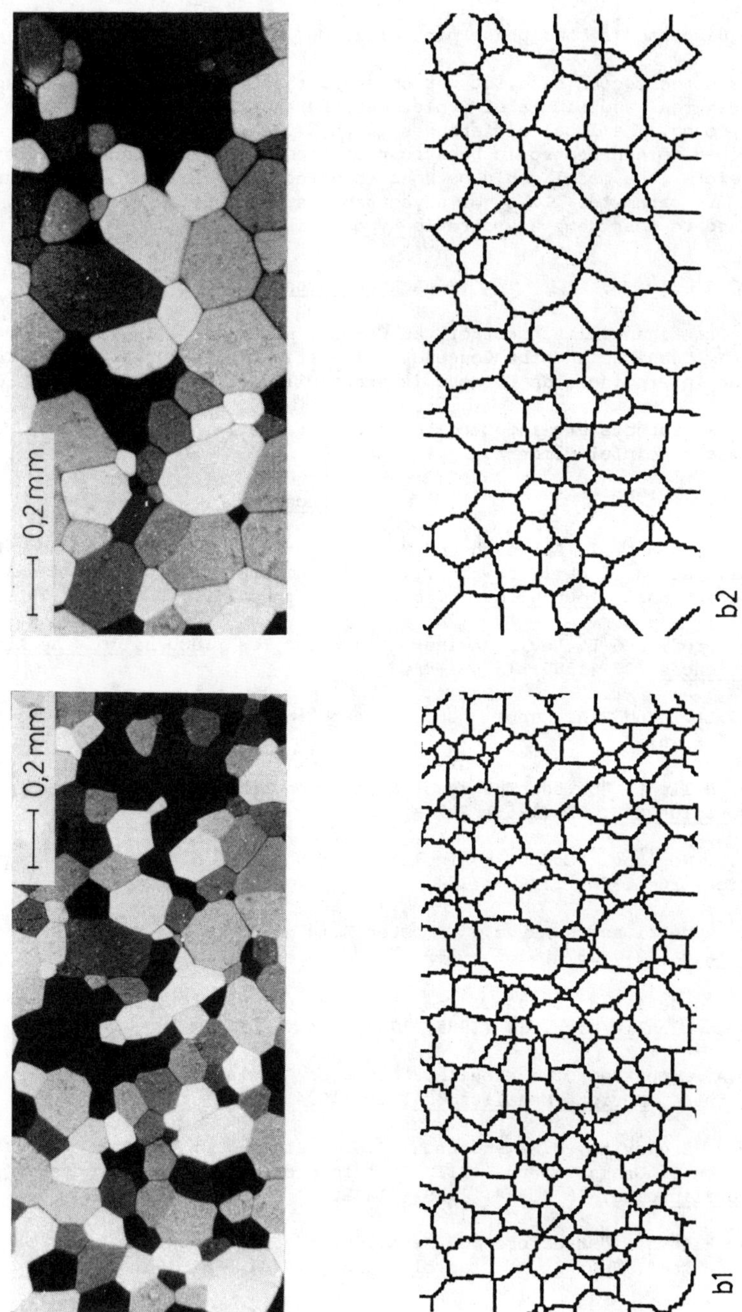

Figure 4 - Comparative evolution of a real and a digitized sample. a) Thin sheet of a polycrystalline ice annealed at - 2 °C; annealing time: a1) 1 hs, a2) 3 hs. b1) Digitization of a1; b2) crystal evolution after 10 simulation steps.

of a dispersed second phase particles.

In the second case, it was observed that there is an average agreement between real and simulated evolutions. Furthermore, adequate values of $M\sigma_{gb}$ were found. The local differences in the evolution found in Fig. 4 could be ascribed to non-isotropic behaviour of the ice grain boundary migration. Therefore, the model could be used to detect non-isotropic behaviour of grin boundary parameter $M\sigma_{gb}$ or anysotropic surface effects. Moreover, it can be used to test some anysotropic hypothesis.

Acknowlegments

The financial support of the Consejo Nacional de Investigaciones Cientí ficas y Técnicas and the Consejo de Investigaciones Científicas y Tecnológicas de la Provincia de Córdoba is gratefully acknowledged.

The authors are indebted to Dra. Laura Levi and to Dr. Jorge M. Caranti for their helpful comments.

References

1. C.A. Knight, T. Ashworth and N.C. Knight, "Cylindrical Ice Accretions as Simulation of Hail Growth: II. The Structure of Fresh and Annealed Accretions", Journal of Atmospheric Science, 35 (1978) 1997-2009.

2. F. Prodi and L. Levi, "Aging of Accreted Ice", Journal of the Atmospheric Sciences, 37 (1980) 1376-1384.

3. L. Levi and E.A. Ceppi, "Grain Growth in Ice", Il Nuovo Cimento, 5C (1982) 445 - 461.

4. M. Hillert, "On the Theory of Normal and Abnormal Grain Growth", Acta Metallurgica, 13 (1965) 227 - 238.

5. V.Y. Novikov, "Computer Simulation of Normal Grain Growth", Acta Metallurgica, 26 (1978) 1739 - 1744.

6. O. Hunderi and N. Ryum, "Computer Simulation of Grain Growth", Acta Metallurgica, 27 (1979) 161 - 165.

7. O. Hunderi and N. Ryum, "Computer Simulation of Stagnation in Grain Growth" Acta Metallurgica, 19 (1981) 1737 -1745.

8. E.A. Ceppi and O.B. Nasello, "Computer Simulation of Bidimensional Grain Growth", Scripta Metallurgica, 18 (1984) 1221 - 1225.

9. D.J. Srolovitz, M.P. Anderson, G.S. Grest and P.S. Sahni, "Computer Simulation of Grain Growth - III. Influence of a Particle Dispersion", Acta Metallurgica, 18 (1984) 1429 - 1438.

10. A. Higashi, "Structure and Behaviour of Grain Boundaries in Polycrystalline Ice", Journal of Glaciology, 21 (1978) 589 - 605.

COMPUTER SIMULATION OF MICROSTRUCTURAL DYNAMICS

G. S. Grest*, M. P. Anderson* and D. J. Srolovitz**

*Corporate Research Science Laboratory
Exxon Research and Engineering Company
Annandale, New Jersey 08801

**Los Alamos National Laboratory
Los Alamos, New Mexico 87545

Abstract

Since many of the physical properties of materials are determined by their microstructure, it is important to be able to predict and control microstructural development. A number of approaches have been taken to study this problem, but they assume that the grains can be described as spherical or hexagonal and that growth occurs in an average environment. We have developed a new technique to bridge the gap between the atomistic interactions and the macroscopic scale by discretizing the continuum system such that the microstructure retains its topological connectedness, yet is amenable to computer simulations. Using this technique we have studied grain growth in polycrystalline aggregates. The temporal evolution and grain morphology of our model are in excellent agreement with experimental results for metals and ceramics.

Introduction

The physical and chemical properties of materials are determined, in part, by their microstructure. Both grain size and texture affect yield strength, fracture, surface adsorption phenomena and other properties. The final grain morphology can be modified by thermal processing, addition of a second phase, deformation, etc. However, in order to effectively tailor the microstructure for specific applications, the mechanism and kinetics of grain growth and recrystallization must be understood. Unfortunately, present theories (1-5) predict grain growth kinetics which differ from experimental observations, have little ability to predict microstructural details and cannot be readily generalized to account for the variety of experimentally controllable features.

Historically, a number of different theoretical approaches have been taken to understand and predict the microstructures of polycrystalline materials. These include analytical theory (1-5), finite element analysis (6), molecular dynamics simulations (7) and physical-analogue models [e.g. bubble rafts] (8), hard sphere models (9) and photoelastic models (10). Whereas analytical models have the virtue of producing closed form solutions, they are usually too oversimplified to quantitatively describe real physical systems. Even when numerical solutions of these analytical theories are performed, this difficulty persists. In principle, molecular dynamics is an ideal way to consistently incorporate the known forces and kinetic processes involved in microstructural evolution. However, present day computer memory and speed limits the technique to small clusters of atoms ($\lesssim 10^5$), which are typically too small to effectively represent more than a single microstructural feature. Whereas physical-analogue models lead to some insight, they are usually crude analogues to the actual physical process.

Since one of the essential features of grain growth or recrystallization in polycrystalline materials involves the competition between individual grains, of different orientations which share a common boundary, it is not surprising that existing grain growth theories have done so poorly in describing more than simple qualitative aspects of the experimentally observed kinetics and topology. Most analytic theories (1,3,4) implicitly assume that grains can be described as spherical, and that growth occurs in an average environment. This type of theory leads immediately to the result that the average grain radius

$$\bar{R} \simeq B\, t^n \tag{1}$$

where B is a temperature dependent constant and $n = 1/2$. This is in contrast to experimental data for n which, while showing substantial scatter, consistently give $n < 1/2$, with a mean value of approximately 0.4 (11). In addition, these macroscopic theories, incorrectly predict the grain size distribution function and make no prediction at all for many topological quantities, like the distribution of the number of grain edges. These failures and inabilities may be traced to their treating grain growth in terms of a grain in a mean environment. This type of mean field theory averages out much of the important, topological effects associated with connectedness of the actual microstructure. Generalizations of this type of theory (12-13) to include many interacting spherical grains, are also not sufficient as they still do not include the simple space-filling requirements which result in a topologically connected structure.

The present outlook for finding a solution of an analytical model for grain growth, which can correctly predict the topological features, does not seem hopeful. In addition, it is also not possible at this time to carry out a realistic atomistic simulation using molecular dynamics to study grain growth or recrystallization. The computer resources for simulating a system of even 10^6 atoms is several years away. Thus, a method to bridge this gap between the atomistic and macroscopic scale is necessary. One method which we (14-17) have pursued is to discretize continuum models for microstructural evolution such that they are amenable to large scale computer simulations. The essential elements of this approach (which will be described in more detail below) consist of mapping the continuum microstructure onto a discrete lattice and defining interactions and dynamics for individual, discrete elements which are analogous to those in continuous systems.

There are several advantages to this type of procedure. The first is that one can incorporate at the most basic level different, often competing, driving forces. This is done by defining the energetics, and hence forces, for each individual element. These energetics include both interactions with surrounding elements and external stimuli. The second advantage is that, not only are complex microstructures generated from simple postulates, but that one can observe the temporal evolution of the microstructures, as well as the micro-mechanisms which lead to its development. In addition, microstructures may be generated which are consistent with any set of postulates as to the nature of the local dynamics and energetics. In this way, the testing of theoretical hypotheses can be performed by comparing simulated microstructures against those experimentally observed, similar to structure determination in x-ray diffraction (18) and lattice imaging (19) in electron microscopy. Finally, in terms of computer resources, both the memory and CPU time requirements are readily available on today's main frame computers.

In this paper we will describe, in more detail, the mapping onto a discrete lattice and the Monte Carlo alogarithm used in the simulations. As an example of how the method works, we will review our work on grain growth. In a companion paper (20) in these Proceedings, we will describe our work on abnormal grain growth and recrystallization. Finally, we will review several possible applications of the methodology to other problems.

The Model and Monte Carlo Method

In order to incorporate the complexity of grain boundary topology, the microstructure is mapped onto a discrete lattice (Fig. 1). Each lattice site is assigned a number between 1 and Q corresponding to the orientation of the grain in which it is embedded. We choose Q large enough so that grains of like orientation impinge infrequently, typically Q = 48 or 64. Since the speed of the computer alogarithm decreases with increasing Q, larger values of Q are not routinely employed. However, for $Q \gtrsim 36$, our results are insensitive to the magnitude of Q. In the present model, a grain boundary segment is defined to lie between two sites of unlike orientation. The grain boundary energy is specified by defining an interaction between lattice sites within a given distance [usually nearest neighbor in two-dimensions (2-d) and up to the third nearest neighbors in three-dimensions (3-d)],

$$E_i = -J \sum_j (\delta_{S_i S_j} - 1). \qquad (2)$$

Here S_i is the orientation of site i ($1 \leq S_i \leq Q$), δ_{AB} is the Kronecker delta and the sum is taken over all sites within a specified distance.

The kinetics of boundary motion are simulated via a Monte Carlo technique in which a site is selected at random and re-oriented to a

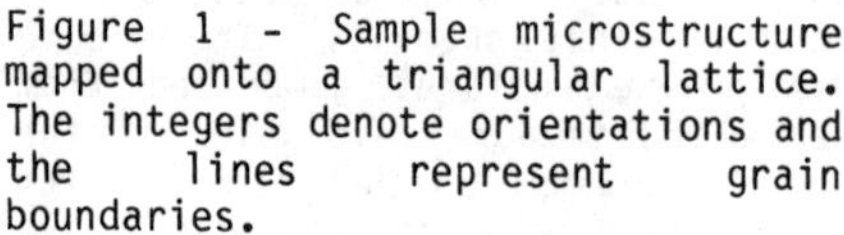

Figure 1 - Sample microstructure mapped onto a triangular lattice. The integers denote orientations and the lines represent grain boundaries.

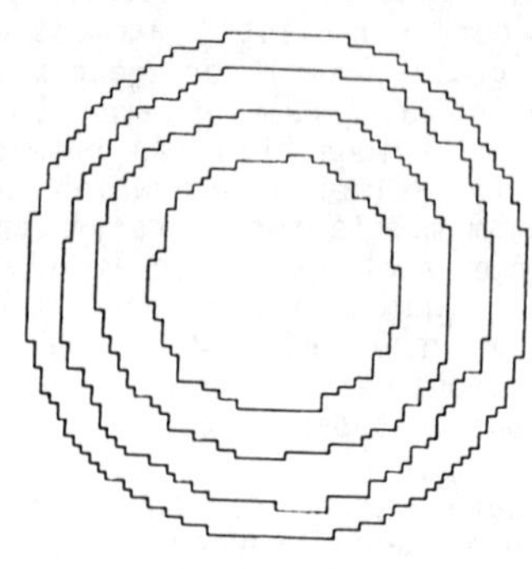

Figure 2 - Evolution of an initially spherical grain of initial radius R_o = 30 (in units of lattice sites) in an infinite matrix on a simple cubic lattice for t = 0, 200, 400 and 600 MCS.

randomly chosen orientation between 1 and Q. If the change in energy associated with the re-orientation, ΔE, is less than or equal to zero the re-orientation is accepted. However, if the re-orientation attempt results in $\Delta E > 0$, the re-orientation is accepted with probability $\exp(-\Delta E/k_BT)$, where k_BT is the thermal energy. A modified version of this alogorithm has been employed to make the simulation more efficient (see Ref. 17). A unit segment of grain boundary, therefore, moves with a velocity, $v_i = C[1-\exp(-\Delta E_i/k_BT)]$, where C is a constant proportional to the boundary mobility. This description of the local boundary velocity is formally equivalent to that derived from classical reaction rate theory. Time, in these simulations, is proportional to the number of re-orientation attempts. N re-orientation attempts is used as the unit of time and is referred to as 1 Monte Carlo Step (MCS), where N is the number of lattice sites. The conversion from MCS to real time has an implicit activation energy factor, $\exp(-W/k_BT)$, which corresponds to the atomic jump frequency. Since the quoted times are normalized by the jump frequency, the only effect of choosing $T \cong 0$ (as in these simulations) is to restrict the accepted re-orientation attempts to those with $\Delta E \leq 0$. Re-orientation of a site at a grain boundary corresponds to boundary migration.

Results

We have carried out extensive simulations in both 2-d and 3-d. Most of our studies in 2-d have been on a triangular lattice of size 200^2, though we have also carried out simulations on the square and honeycomb lattice. Our results for 3-d are for the simple cubic lattice of size 60^3 and 100^3. One remarkable conclusion from our study is that most of the results are independent of dimensionality (for $d \geq 2$). For example, the grain area distribution and topological distribution of the number of edges is the same for the growth in the plane and for the cross-section of the 3-d microstructures. This result allows us to carry out many simulations in two-dimensions, where it is possible to study systems with much larger linear dimensions than in 3-d.

As a test of the simulation procedure, it is useful to examine the simple case of a circular or spherical grain shrinking in an infinite

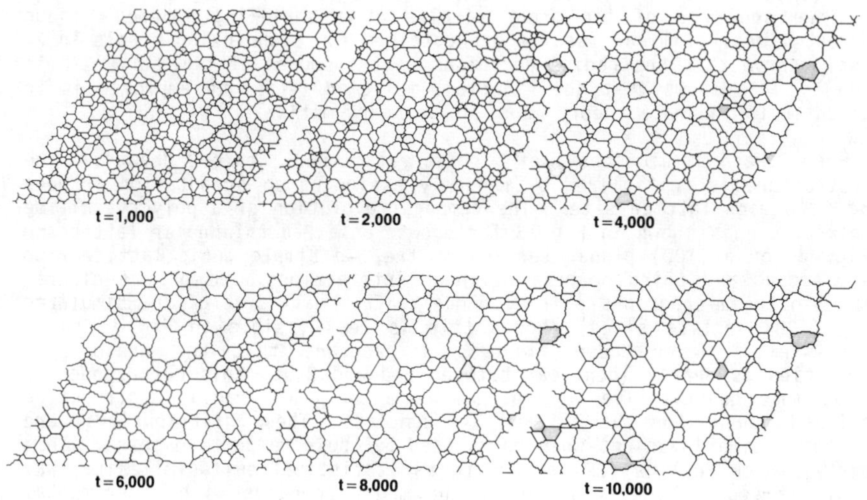

Figure 3 - Temporal evolution of a two-dimensional polycrystalline microstructure on a 200^2 triangular lattice for a pure, isotropic material. Grains with six edges have been shaded.

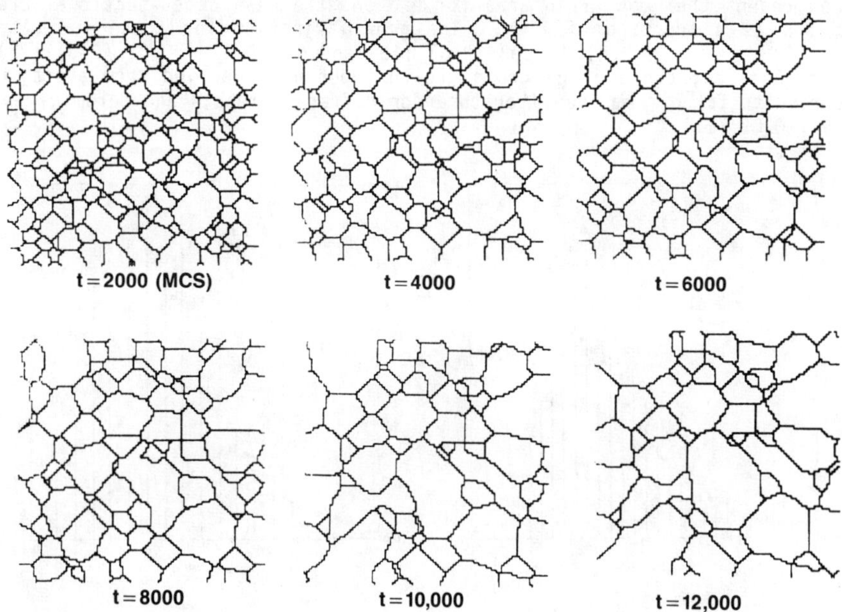

Figure 4 - Temporal evolution of the cross-section from the 3-d microstructure for Q = 48 on a 100^3 simple cubic lattice.

matrix (21). This is the case where the analytical theory applies. In Fig. 2 we show one of the great circles of a shrinking sphere at four different times. Similar results are obtained for a shrinking circle in 2-d. We observe that the radius decreases as the square of time, $n = 1/2$ in Eq. (1). This shows that the kinetic simulation technique employed is in agreement with the rate theory model of boundary motion.

For the simulations of polycrystalline grain growth, the microstructure is initialized by randomly assigning an orientation between 1 and Q to each lattice site. The temporal evolution of a polycrystalline microstructure is shown in Fig. 3 for growth on a 2-d triangular lattic and in Fig. 4 for a (100) planar section on the 3-d simple cubic lattice with first, second and third nearest neighbor interactions. Steps are clearly visible in the micrographs in Fig. 4 due to the small size of the simulated volume (100^3 lattice sites) though this is less visible in Fig. 3 for a 200^2 system. From these figures, it appears that the grain size distribution is nearly identical between 2-d and 3-d. This correspondence is made more quantitative in Fig. 5, where the grain radius distribution function is shown for the 2-d and 3-d samples. These distributions were found to be time invariant when normalized by their respective means. This property, which Mullins (22) refers to as statistical self-similarity, was always observed. To test how well our model compares with experimental data, we present the cross-sectional area grain size distribution taken from data like that presented in Fig. 4 and compared with that measured for pure Fe in Fig. 6. Note that the grain size distribution function for the simulations and experiment agree remarkably well.

The mean grain size was also monitored as a function of time. In Fig. 7 we present the mean grain area for 2-d and the mean cross-sectional area for 3-d as a function of time. The growth kinetics for these two cases gave ($A \sim t^{2n}$) $n = 0.41 \pm 0.03$ (2-d) (15) and $n = 0.37 \pm 0.02$ (3-d) (23). For comparison, the average grain growth exponent n derived from averaging the results for n from the literature for different metals and ceramics is 0.39 ± 0.07 (11).

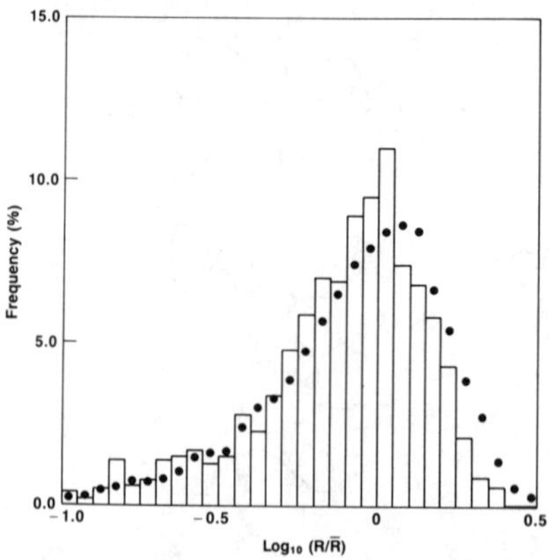

Figure 5 - Grain radius distribution as determined from cross-sections of the 3-d lattice model (filled circles) and from the 2-d lattice model (histogram).

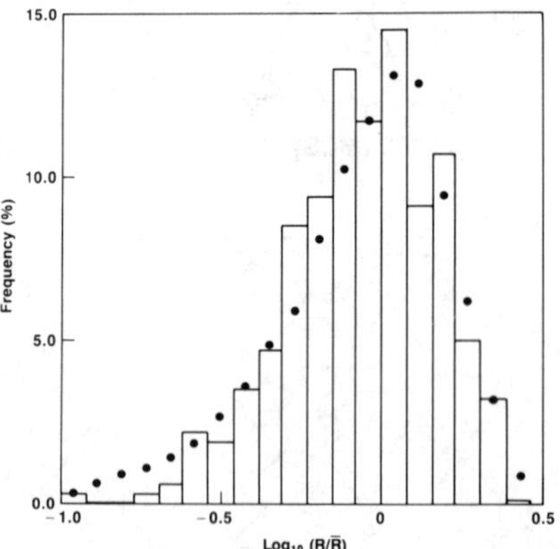

Figure 6 - Grain radius distribution as determined from a cross-sectional area analysis of pure Fe (histogram) and from cross-sections of the three-dimensional lattice model (filled circles).

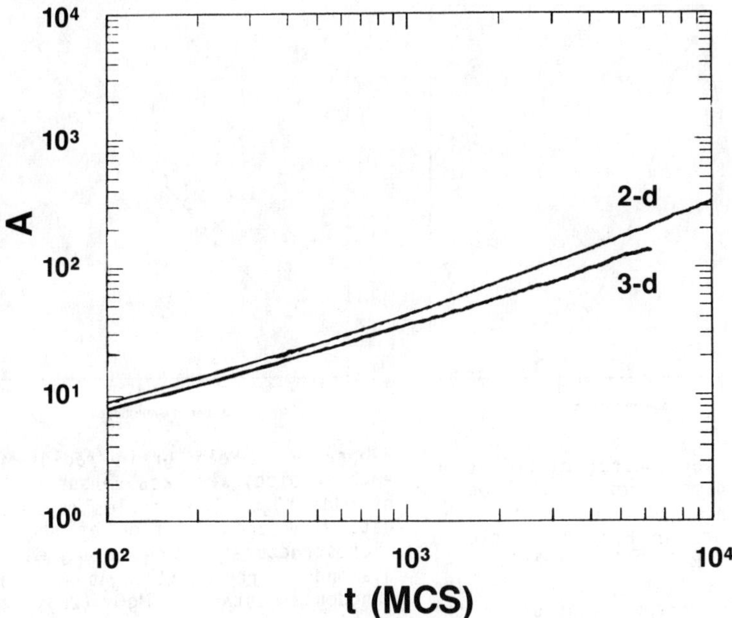

Figure 7 - Log-log plot of the mean area A versus time for Q = 48 for the two-dimensional grain growth simulations and for the cross-sectional data from the three-dimensional simulations.

One advantage of this method is that since grain topology is not averaged away, we can measure the topological distribution functions as shown in Fig. 8. Here we compare the results from our simulation in 3-d with those measured for Al (24), Sn (1), and MgO (25). In Fig. 9 we show the mean grain radius for each topological class versus the number of edges, N_e. In all cases, the agreement between the simulations of our lattice model and the experimental results is excellent.

One interesting question that often arises is why the growth exponent $n < 1/2$ in both our simulations and experiments. Recently, Mullins (22) has shown that by assuming statistical self-similarity (which our simulations demonstrate) and local equilibrium that n must equal 1/2 in 2-d. He uses the result, first proven by von Neumann (26) and Mullins (27), that for both bubble growth and idealized grain growth in 2-d, the rate of change of the area A_i of an individual grain (or bubble) depends only on the number of sides,

$$\dot{A} \equiv \frac{dA_i}{dt} = \frac{\pi k}{3}(N_e - 6) \tag{3}$$

where k is a constant. This equation should be valid for an arbitrarily shaped two-dimensional grain of N_e-sides under the assumption that $u = k/R$ and the angle between intersecting grain boundaries is 120°. Here u is the local velocity and R is the signed local radius of curvature lying in the plane, counted positive when it lies along n . Thus $\dot{A}_i > 0$ for $N_e > 6$ and is < 0 for $N_e < 6$. Writing the statistical self-similarity hypothesis in the form

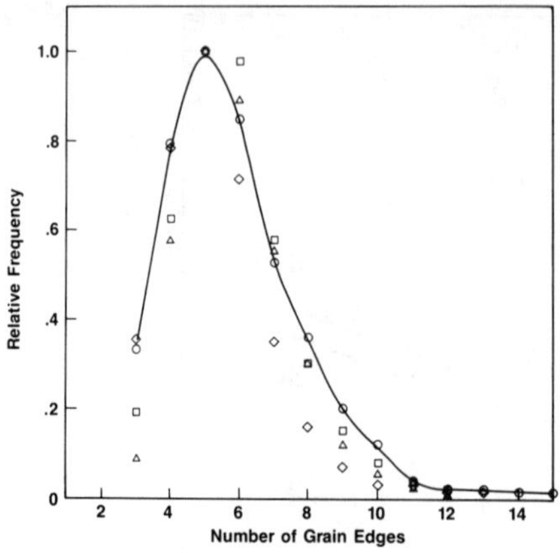

Figure 8 - Topological distribution for the number of grain edges measured in cross-section for the microstructure in Fig. 4 (circles), and for MgO (triangle), Al (squares), and Sn (diamond). The curve is drawn through the simulation results.

Figure 9 - Mean grain radius for each topological class versus number of edges N_e. The circles represent data from cross-section of the 3-d microstructures. The triangles and diamonds represent Aboav and Langdon's data on MgO (25), and Beck's data on Al (24), respectively. The squares represent the results obtained from the 2-d simulation. The line is drawn through the 3-d simulation results.

$$f_{N_e}(A,t) = \phi_{N_e}(A/\bar{A})/\bar{A} \qquad (4)$$

where $f_{N_e}(A,t)dA$ is the probability that a grain has N_e edges and area between A and A + dA, Mullins (22) then proves that if local equilibrium is satisfied [Eq. (3) is valid for all grains at all times] then n = 1/2.

Since we have already shown (16) that Eq. (4) is valid, it is of interest to test Eq. (3). We first performed a grain growth simulation in 2-d for 1,000 MCS until there were 960 grains remaining. We then followed each grain, monitoring its number of edges as a function of time. In Fig. 10, we present results for the temporal evolution of the area of several individual grains with the same number of edges. Those grains monitored in Fig. 10 were highly unusual in that their number of edges did not change during the majority of the 10,000 MCS step run. For convenience we have reset the clock to 0 after we started monitoring the number of edges and we have plotted a symbol only when the number of edges equal the number N_e specified in the figure. In Fig. 4, we have shaded those grains which remain predominantly 6 edged during the 10,000 MCS run. From Fig. 10, we see that the slope of $\dot{A}_i$ is approximately zero for N_e = 6 and that for N_e = 4 is approximately of equal magnitude and opposite sign from that for N_e = 8. An average slope for N_e = 5 and 7 is harder to determine. We also must point out, however, that most of the grains change N_e too often to be plotted in this way. Shown in Fig. 11 is the temporal evolution of the area, A_i, for five more typical grains. A different symbol is used to label the number of edges at a given time. Note how often N_e changes. The data in Fig. 11 are the norm, while those in Fig. 10, are the exceptions.

Thus it appears that for those grains which keep the same number of sides for a long time, Eq. (3) is satisfied. However, there are large fluctuations around the mean slope and for most grains Eq. (3) is not applicable. From these results, we can understand the discrepancy in the value of n between classical theories (1-5) and our simulations. Since local equilibrium, Eq. (3), is never established, the grain growth is slower than would be expected. The grains are continuously changing their number of edges N_e too rapidly for the system to reach local equilibrium.

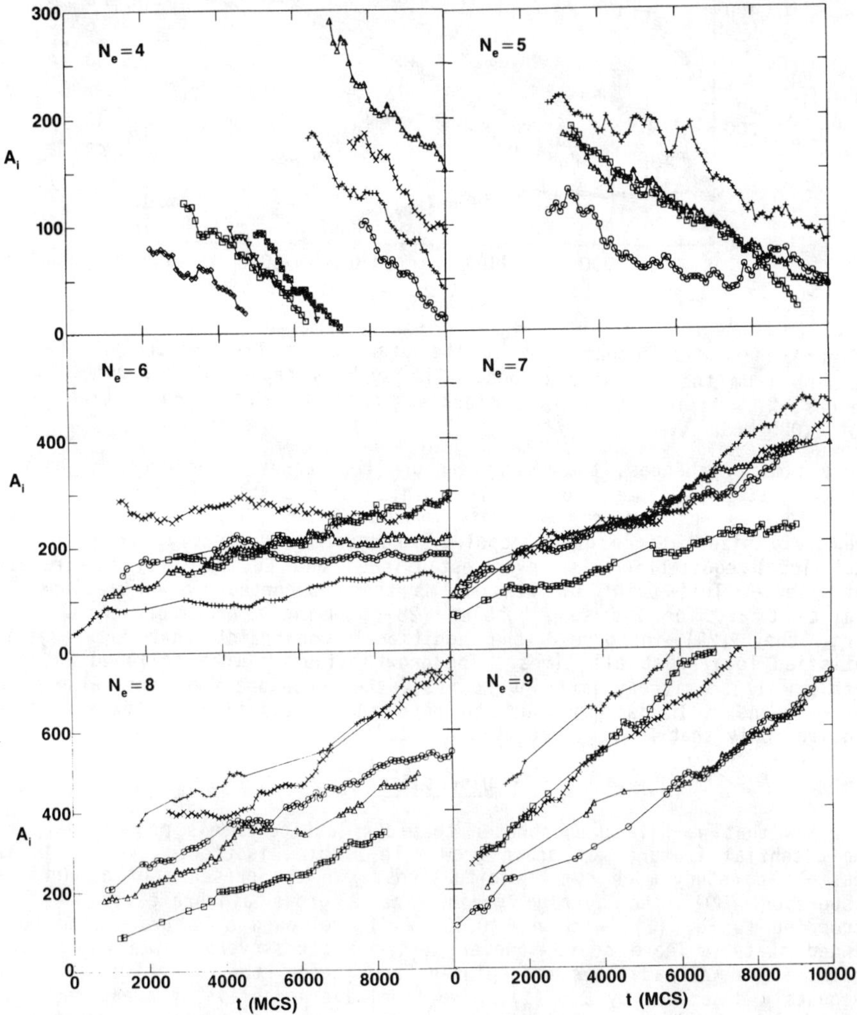

Figure 10 - Time dependence of the area, A_i, of individual grains from the 2-d simulations with a specific number of edges, N_e. Symbols are shown every 20 MCS when the grain has the number of edges specified. The simulation was started from a random starting state and run for 1,000 MCS, after which the clock was reset to 0. The data plotted are for grains which had the specified number of edges over unusually large portions of the 10,000 MCS simulation.

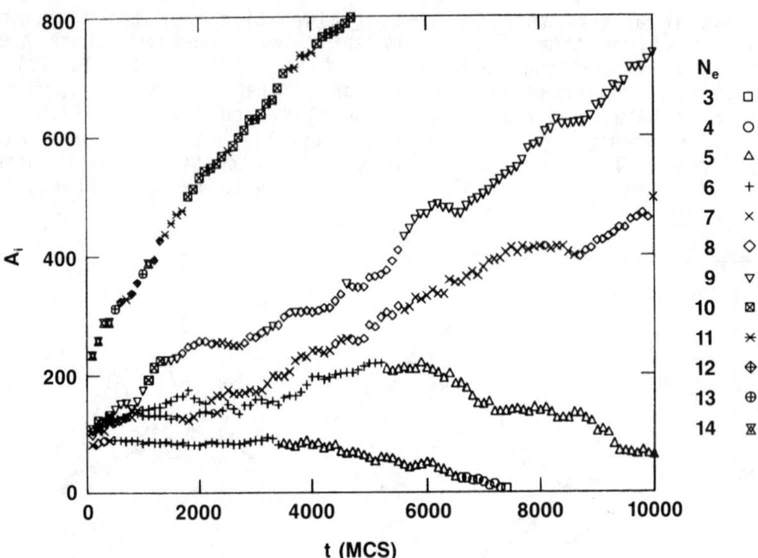

Figure 11 - Time dependence of the area A_i of five randomly chosen grains from the 2-d simulations. The symbols represent the number of edges, N_e. As in Fig. 10, the clock was reset to 0 after an initial run of 1,000 MCS.

Every time N_e changes, the microstructure rearranges to accomodate the new growth rate, as per Eq. (3). This accomodation process is not instantaneous but requires a finite amount of time, τ , which scales with the grain size. Since grain topology changes so frequently, it is likely that local equilibrium is never established and Eq. (3) is only rarely satisfied. This point of view is further supported by simulations of Weaire et al. for 2-d soap bubbles (28). Using a procedure similar to ours, they (29) introduced the additional constraint that Eq. (3) be satisfied locally at all times. The growth kinetics then followed Eq. (1) with n = 1/2. Without imposing Eq. (3), they obtained the same value of n as we found. It is important to note that statistical self-similarity requires only that $R \sim t^n$, not that n = 1/2!

Discussion

Now that we have demonstrated that our lattice models produce many of the essential features of grain growth in simple, isotropic system, it is possible to study more complex situations. In the presence of a particle dispersion (30), the average grain size R grows algebraically in time according to Eq. (1), with $n \sim 0.40$ (2-d), followed by a transition to a pinned state. There does, however, exist a cross-over regime where some grains have already becomed pinned and hence the kinetics are not adequately described by Eq. (1). The final average grain area and the time required for the microstructure to pin are both approximately proportional to the inverse particle concentration in 2-d. In the presence of anisotropic grain boundary energies (31), the growth exponent n is found to decrease in 2-d from n = 0.42 ± 0.02 to 0.25 ± 0.02 and the grain size distribution broadens, as the anisotropy is increased. However, for values of the anisotropy expected for real polycrystalline materials, the results differ only slightly from those measured in the isotropic case. In addition, abnormal grain growth (32) and recrystallization (33) have also been studied. Results from this work will be presented in these proceedings (20).

We think that we have demonstrated that although the lattice model employed in the simulation described above is inherently discrete, the resultant micrographs show an excellent correspondence with those from real materials. The constraints imposed upon the boundary intersections due to the symmetry of the underlying lattice appears to be unimportant, and the interface tension dictates the angle at which grain boundaries meet. What is somewhat surprising is the agreement between the two-dimensional results and the cross-sections of the three-dimensional systems. While the grain radii distribution function as computed from the volume is narrower than the grain radii taken from the cross-sectional area, the grain radii taken from the cross-sectional area in 3-d are nearly identical to those found in the 2-d simulations.

One further test of the validity of the lattice model is the comparison of kinetic predictions with experiment. Kinetics measured from the mean chord, area and volume give the same growth exponent, n, in 3-d. While we do not know if the difference in n between 2-d and 3-d is significant, both results agree very well with experimental findings and disagree with existing grain growth theories (n=1/2). While one cannot rule out that asymptotically both the simulations and experiments would not eventually cross over to n = 1/2, we see no physical reason why the growth of a grain in an average environment should be the same as that occurring in the presence of many, competing grains. In addition, the fact that both the kinetics and all of the measured grain size distributions agree so well with experiment, suggests that the model has captured the essential features of the grain growth, at least on the time scale of an experimentalist.

REFERENCES

1. P. Feltham, Acta Metall. 5, 97 (1957).
2. F. N. Rhines and K. R. Craig, Met. Trans. 5, 413 (1974).
3. M. Hillert, Acta Metall. 13, 227 (1965).
4. N. P. Louat, Acta Metall. 22, 721 (1974).
5. S. K. Kurtz and F. M. A. Carpay, J. Appl. Phys. 51, 5725 (1981).
6. R. D. Cook, Concepts and Applications of Finite Element Analysis, Wiley, New York (1974).
7. J. Kushick and B. J. Berne, in Statistical Mechanics B; Time Dependent Processes (editor B. J. Berne) Plenum, New York (1977), p. 41.
8. W. L. Bragg and J. F. Nye, Proc. Royal Soc. A190, 474 (1977).
9. J. L. Finney, Ph.D. Thesis, University of London, 1958.
10. J. W. Dally and W. F. Riley, Experimental Stress Analysis, McGraw-Hill, New York (1978).
11. F. Haessner, Recrystallization of Metallic Materials, edited by F. Haessner, Riederer-Verlay, Stuttgart (1978), p. 63.
12. O. Hunderi, N. Ryum and H. Westengen, Acta Metall. 27, 161 (1979); 29, 1737 (1981).
13. V. Yu Novikov, Acta Metall. 26, 1739 (1978); 27, 1461, (1979).
14. D. J. Srolovitz, M. P. Anderson, G. S. Grest and P. S. Sahni, Scripta Metall. 17, 241 (1983).
15. M. P. Anderson, D. J. Srolovitz, G. S. Grest and P. S. Sahni, Acta Metall. 32, 783 (1984).
16. D. J. Srolovitz, M. P. Anderson, P. S. Sahni and G. S. Grest, Acta Metall. 32, 793 (1984).
17. P. S. Sahni, D. J. Srolovitz, G. S. Grest, M. P. Anderson and S. A.

Safran, Phys. Rev. B28, 2705 (1983).
18. M. J. Buerger, X-Ray Crystallography, Wiley, New York (1942).
19. J. A. Venables, Developments in Electron Microscopy and Analysis, Academic Press, New York (1976).
20. M. P. Anderson, G. S. Grest and D. J. Srolovitz, This Proceedings
21. G. S. Grest, S. A. Safran and P. S. Sahni, J. Appl. Phys. 55, 2432 (1984).
22. W. W. Mullins (to be published).
23. M. P. Anderson, G. S. Grest, and D. J. Srolovitz, Scripta Metall. 19, 225 (1985).
24. P. A. Beck, Phil. Mag., Suppl. 3, 245 (1954).
25. D. A. Aboav and T. G. Langdon, Metallography 1, 333 (1969).
26. J. von Neumann, in Metal Interfaces, p. 108, ASM, Metals Park, Ohio (1952).
27. W. W. Mullins, J. Appl. Phys. 27, 900 (1956).
28. D. Weaire and J. P. Kermode, Phil Mag. 50, 379 (1984).
29. D. Weaire, private communication.
30. D. J. Srolovitz, M. P. Anderson, G. S. Grest and P. S. Sahni, Acta. Metall. 32, 1429 (1984).
31. G. S. Grest, M. P. Anderson and D. J. Srolovitz, Acta Metall. 33, 509 (1985).
32. D. J. Srolovitz, G. S. Grest and M. P. Anderson, Acta Metall., in press.
33. D. J. Srolovitz, G. S. Grest and M. P. Anderson, submitted to Acta Metall.

MICROSTRUCTURAL EVOLUTION IN THIN FILMS

H.J. Frost* and C.V.Thompson**

*Thayer School of Engineering
Dartmouth College
Hanover, NH 03755

**Department of Materials Science and Engineering
Massachusetts Institute of Technology
Cambridge, MA 02139

Abstract

The geometry and topology of two dimensional grain structures have been analyzed for different conditions of nucleation and growth to impingement. These conditions include simultaneous nucleation at a fixed number of sites, and continuous nucleation at a constant rate per unit of untransformed area. These cases result in markedly different structures.

The grain boundary migration that occurs after impingement has also been modelled by a program which moves boundary segments and triple points according to the capillary forces due to the curvatures. Preliminary results will be presented. This modelling should have direct applications to thin films of metals or semiconductors.

Introduction

In this paper we present simple models for the development of microstructures in polycrystalline thin films. The processes modelled are crystal nucleation and growth to impingement, followed by grain boundary motion. These processes determine the topology and geometry of the resulting grain structure. Modelling of film formation provides: i) tools for post-formation analysis of formation conditions, ii) a framework for analyzing the impact of modified formation conditions on resulting microstructure, and iii) initial microstructures for subsequent modelling of grain growth. Our modelling studies are coupled with ongoing experimental research on microstructure and microstructural evolution in thin films of metals and semiconductors.

A thin film can form by deposition from a vapor or liquid onto a substrate, or by crystallization of an amorphous or liquid film. We will limit this discussion to the formation of a planar film on a planar substrate. In the case of film deposition, crystallization proceeds via crystal nucleation, normally at a number of locations, followed by crystal growth. Nucleated crystals grow until they impinge on one another, leading to grain boundaries. Under most conditions of interest, crystal nucleation occurs at the substrate surface or, in the context of this paper, within a plane. Nucleation within a plane can occur homogeneously, at random locations, or heterogeneously, at a limited number of heterogeneities in the substrate surface. The rate of nucleation (per unit area) can vary from constant, as is normally the case for homogeneous nucleation, to decaying with time, as is normally the case for heterogeneous nucleation. Crystallization of an amorphous or liquid film is also likely to initiate through nucleation at a surface, and is therefore phenomenologically identical to nucleation during deposition.

We treat two extreme nucleation conditions: continuous nucleation and nucleation-site saturation. In the case of continuous nucleation, we assume that nucleation occurs randomly in a plane and that the nucleation rate remains constant throughout the process of film formation. In the case of site saturation, we assume that there is a limited number of randomly distributed sites at which nucleation occurs simultaneously. We also consider an intermediate case in which growing crystals are surrounded by a region, of thickness d, in which nucleation cannot occur. This case is most relevant to film formation by deposition. During deposition it is likely that regions adjacent to growing crystals are sufficiently depleted of monomers that nucleation cannot occur. The width of this depletion region is a function of the rate of crystal growth and the rate at which monomers are resupplied at the surface. We have assumed that the width remains constant with time as the crystal grows, but have modelled various widths so that the impact of this variable on the final grain topology can be described.

In all cases, we assume a constant and isotropic crystal growth rate. The computer model to be described could, however, be readily modified to account for growth velocity anisotropy and non-random orientations among nuclei. Variable growth rates could also be modelled.

When film formation occurs under conditions for which grain boundaries are mobile, the final grain topology will be modified. We have developed the means of modelling grain boundary motion and grain growth, and here illustrate the use of this technique on a small sample array of grains. Experimental results suggest that under a broad range of conditions, mobile boundaries play an important role in determining the topology of grain structures in as-deposited as well as annealed films.

Calculations

Let us review the geometry that results from nucleation and growth to impingement. If two neighboring grains nucleate at the same time, and grow in radius at the same constant rate, then the boundary at which they meet will be a straight line which is the perpendicular bisector of the line joining the two nucleation sites. This case is illustrated in Figure 1a, which shows the grain shapes at different times. Three neighboring grains that nucleate at the same time and grow at the same rate will meet at a triple point which is the meeting point of the three perpendicular bisectors of the edges of the triangle formed by the sites. This triple point is equidistant from the three nucleation sites, and is therefore the center of the circle specified by the three nucleation sites.

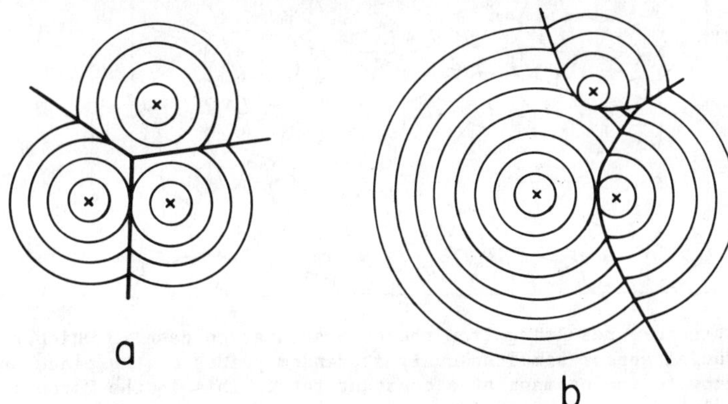

Figure 1: Boundaries formed by the impingement of growing circular grains. The grain edges at successive times are shown as concentric circles. (a) If grains nucleate simultaneously, the boundaries are the perpendicular bisectors between nucleation sites. (b) The boundaries are hyperbolae.

For any array of points, simultaneous nucleation and constant growth rate will result in the well-studied Voronoi polygon structure. The Voronoi polygon for a particular nucleation point is that region of the plane that is closer to that point than to any other nucleation point. It is also known as a Wigner-Seitz cell, a Dirichlet region, or a Theissen polygon. Studies of this structure are reviewed by Getis and Boots (1). An example of the structure is shown in Figure 2. Each grain is the Voronoi polygon for its nucleation point. For the structure in Figure 2, the array of nucleation sites is chosen randomly. Non-random nucleation sites would produce a different Voronoi polygon structure. This two-dimensional site saturation model is similar, but not identical, to the cellular model of Mahin, Hanson, and Morris (2), who reported on planar sections through a three-dimensional model of nucleation and growth to impingement.

If two neighboring grains nucleate at different times, but grow at the same rate, they will impinge along a boundary which forms a hyperbola. Each point on the boundary will be closer to the second grain nucleated than the first grain nucleated by a distance equal to the difference in nucleation times multiplied by the growth rate. Three grains that nucleate at different times will grow to meet at a triple point, as shown in Figure 1b, which is specified by the equation:

$$r_1 + G \times t_1 = r_2 + G \times t_2 = r_3 + G \times t_3 \qquad (1)$$

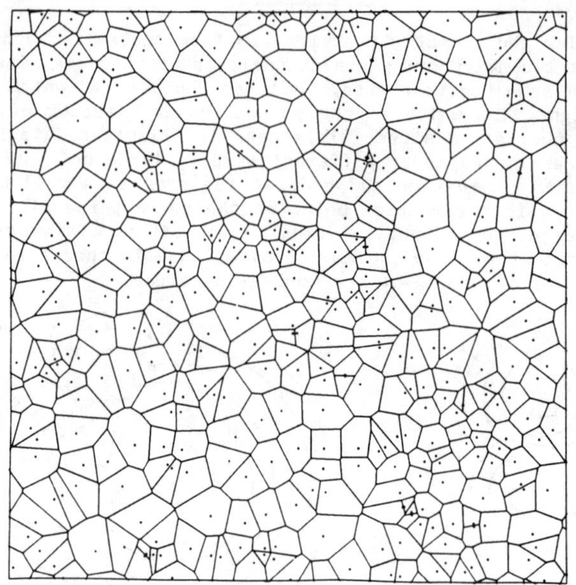

Figure 2: Structure resulting from the site saturation case in which all nuclei appear simultaneously at random points on the plane and grow to impingement at a constant rate. This is the Voronoi polygon construction for random points.

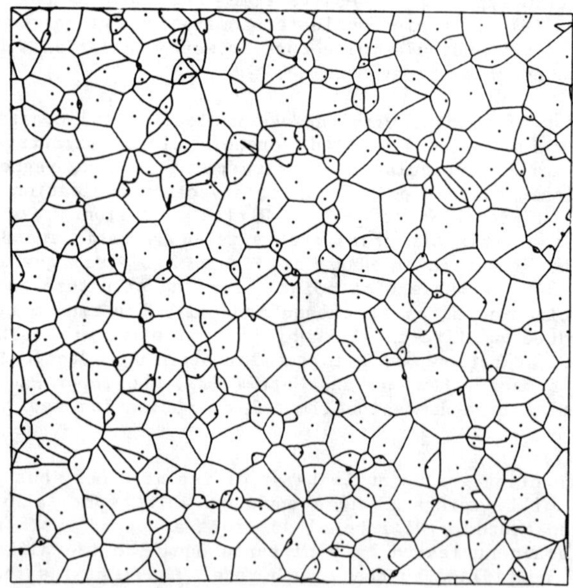

Figure 3: Structure resulting from the Continuous Nucleation case or Johnson and Mehl model in which nucleation continues at a constant rate per unit of untransformed area.

where G is the growth rate, t_i is the nucleation time for grain i, and r_i is the distance from the nucleation site of grain i to the triple point.

We must note that three nearby grains nucleating at different times, producing hyperbolic boundaries, may form one, two, or zero triple points depending on nucleation times and site positions. That is, the hyperbola describing the boundary between grains 1 and 3 may intersect the hyperbola describing the boundary between grains 1 and 2 at zero, one, or two points. The possibility of forming two triple points is an important distinction of the case of different nucleation times (continuous nucleation). It represents the possibility of forming lenses defined by two hyperbolic boundaries. We shall return to this point later.

The structure that results from continuous nucleation, with constant growth rate, was first described in 1939 by Johnson and Mehl (3), and in 1945 by Evans (4). It has been less thoroughly studied than the Voronoi polygon structure, but various properties have been calculated by Meijering (5) and by Gilbert (6). An example of this structure is shown in Figure 3. It is similar, but not identical, to the planar sections through the three-dimensional continuous nucleation of model of Mahin, et al. (2).

The calculation of the continuous nucleation structure begins with picking nucleation sites randomly on a plane at random times. The nucleation sites must be restricted to those areas not already occupied by growing grains. The time variation is created by incrementing the time by random intervals which simulate a Poisson noise distribution. The average nucleation rate is one grain per unit time per unit initial area.

For each time interval a random site is picked. If there is already another grain that has grown to cover that site, then that site is neglected, and another time interval is taken. If there is not already another grain there, then that site adds another growing grain to the list. At this stage we need not worry about the geometry of impingement. Eventually the entire plane is covered by grains, and we will cease to add new grains to the list. The total number of grains nucleated will depend on the ratio of the nucleation rate per unit area and the growth rate. If the growth rate is higher, the plane will be covered more quickly and fewer grains will nucleate.

Given that the structure is random, the nature of the structure does not change with the ratio of nucleation rate to growth rate. The average grain size changes, but the structures for large and small grain sizes are equivalent in topology and differ only by a scaling factor. The average grain area that results from this algorithm is given by Gilbert (6) as:

$$A = 1.1371\ (G/\dot{N})2/3$$

where $\dot{N}$ is the nucleation rate per unit area per unit time, and G is the growth rate. We used a growth rate of 0.82467 units of length per unit time, which, given our nucleation rate, produces an average grain area of 1.0.

Once the nucleation sites and times are calculated for all grains, the triple points and boundary segments are located. This is done by considering all triplets of sites that are nearby each other. The potential triple points are found by finding where the hyperbola boundary between the first and second grains meets the boundary between the first and third grains (at zero, one or two places). This is equivalent to find the point (or points) that satisfy equation (1). Each triple point is formed at a particular time: $t = r_1/G + t_1$. A search is then made over all other nearby grains to determine whether another grain would have reached the potential triple point before the triplet of grains would meet. If another grain did get there first, the potential triple point is not real and is neglected.

Once all triple points are found, the boundary segments can be drawn by calculating the hyperbola boundary location at several points. Since it is possible to form lenses, two neighboring grains may share more than one boundary segment, on either side of a lens. This means that care must be taken in sorting out which triple points bound which boundary segments. The areas of the grains are calculated from the triangles formed by the points used to draw the boundaries and the nucleation site.

In order to model the effect of a depletion region in which nucleation is excluded, we have simply amended the continuous nucleation calculation to exclude any nucleation event that is within a distance δ of the edge of any previously nucleated grain (Figure 4). Since we kept the same nucleation rate (1.0 per unit time per unit area) and the same growth rate (0.8247), we find fewer nucleation events per unit area, and therefore a larger average area. We here present data for two depletion region thicknesses: $\delta = 0.2$, and $\delta = 0.4$. That is, $\delta = 0.2$, or 0.4 of the square root of the average area of grains in the Johnson and Mehl ($\delta = 0$) model. The structures that result are shown in Figure 5.

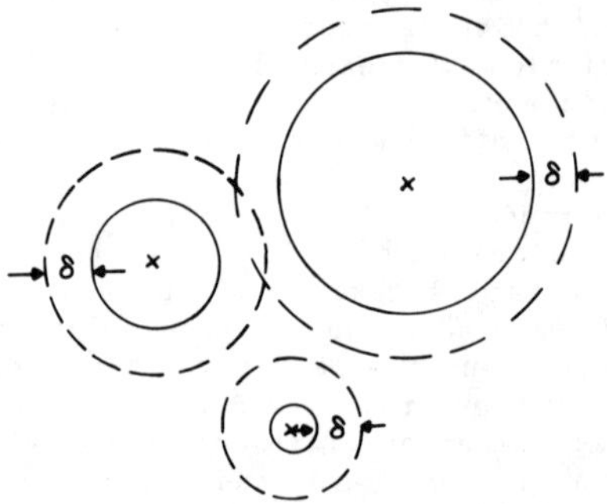

Figure 4: Schematic of the depletion region, or nucleation-exclusion zone, of width δ, which is assumed to surround the growing grains.

For all conditions, the edge effects in the calculations are handled simply by using only those grains whose nucleation sites are within an interior region which has its boundary four grain diameters inside of the outer boundary of the region in which grain nucleation is calculated.

Results

The two different nucleation conditions produce noticeably different structures. The differences go beyond the difference between straight boundaries and hyperbolic boundaries. The structures also differ more or less in almost all their measured topological and geometric properties. Actual microstructures of thin films differ from either of these extremes. The microstructures shown in Figure 5 represent intermediate microstructure resulting from a depletion region in which nucleation is forbidden. These structures more closely resemble thin film microstructures. If grain boundaries are mobile during crystallization or film formation (after grain impingement) altered microstructures will result. In this section we report only on the initial structures, and will consider the effects of boundary migration later.

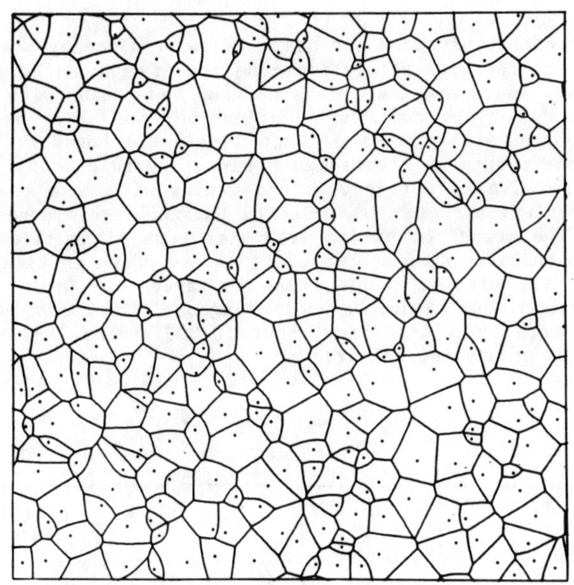

Figure 5a: Structure resulting from the continuous nucleation case when a nucleation-exclusion zone of width $\delta = 0.2$ is placed around each growing grain.

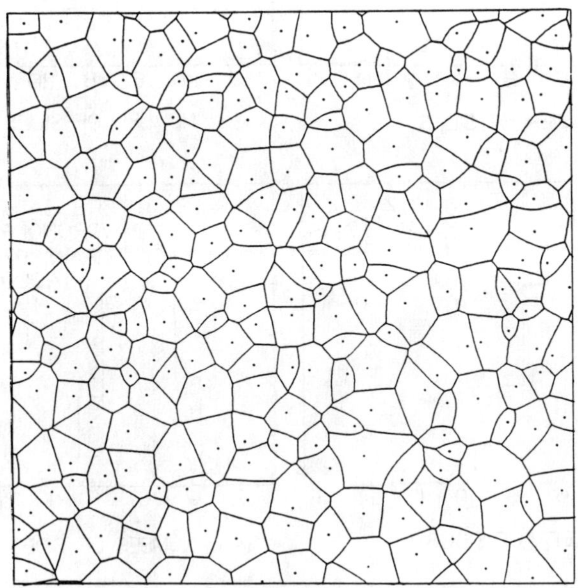

Figure 5b: Structure resulting from the continuous nucleation case when a nucleation-exclusion zone of width $\delta = 0.4$ is placed around each growing grain.

Sides Per Grain

The average number of sides per grain is constrained by planar topology to be six. For the site saturation case the grains can have no fewer than three sides. There are few triangles, and few grains with more than than eight sides. For the continuous nucleation case there are a few lenses which have only two sides. (Each hyperbolic boundary segment between two triple points is treated as a side, regardless of its curvature.) There are also many triangles and grains with more than eight sides. When a depletion region is included, the number of lenses decreases dramatically. The number of grains with more than eight sides is also reduced. The distributions of number of sides are shown in Figure 6. The distribution for the site saturation case is taken from Crain (7), who calculated it from 57,000 random Voronoi polygons. The distributions for the other cases were calculated from about 10,000 grains, and therefore have slightly greater statistical errors.

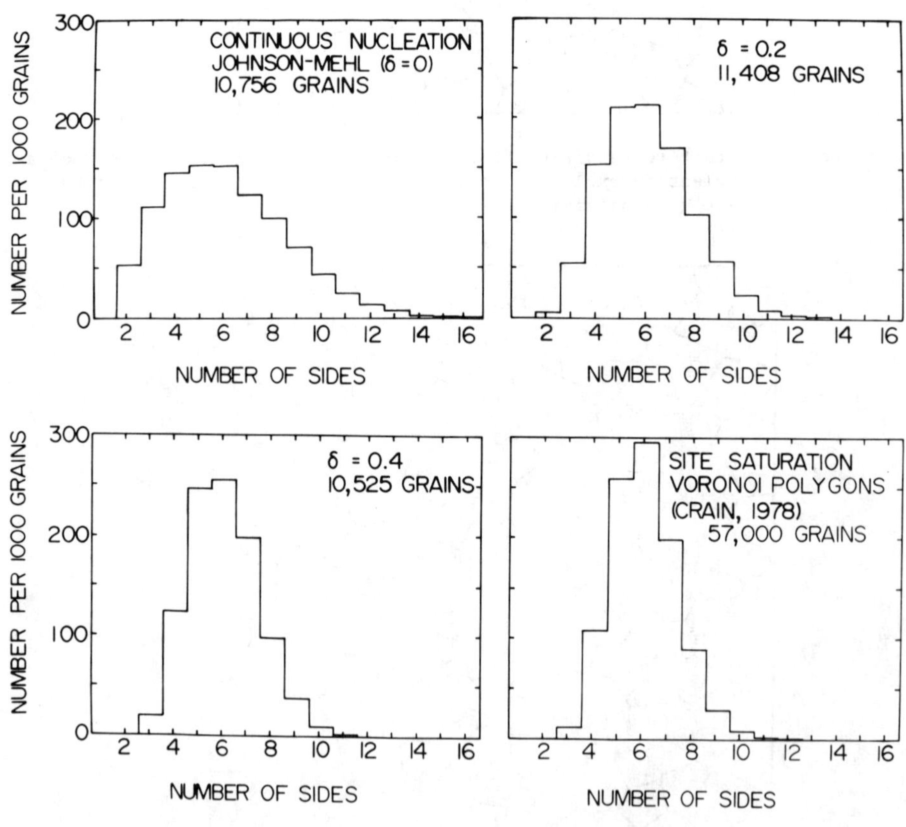

Figure 6: Distributions of the number of sides per grain for the different nucleation conditions.

Grain Size Distribution

The difference between the various cases is shown more dramatically by the grain size distributions shown in Figures 7 and 8. Once again, the distribution for the site saturation case is taken from Crain (7), for 46,000 random Voronoi polygons. The other distributions were determined from our present results from about 10,000 grains. The continuous nucleation case produces far more small grains (mostly lenses and three-sided grains), and more large grains. This is to be expected because in this case the first grains to nucleate should be able to grow larger before impingement. The small grains are those which nucleate last, when there is little free area remaining for them to grow into.

We have not derived an analytic expression for the area distribution for the continuous nucleation case, although that should be possible. We may note, however, that the distribution is similar to an exponential decay. The intercept at A = 0 for the continuous nucleation case is clearly non-zero, while for the other cases it appears to be zero. We may also note that the area distribution for the continuous nucleation case is similar to that calculated for a planar section through a three dimensional Johnson-Mehl model by Mahin et al. (2).

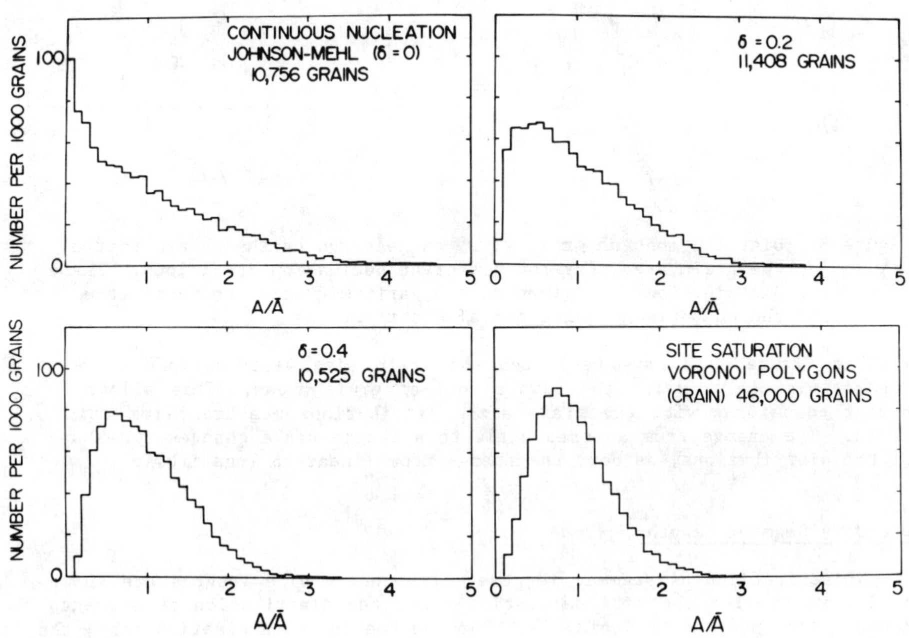

Figure 7: Distributions of grain areas for the different nucleation conditions.

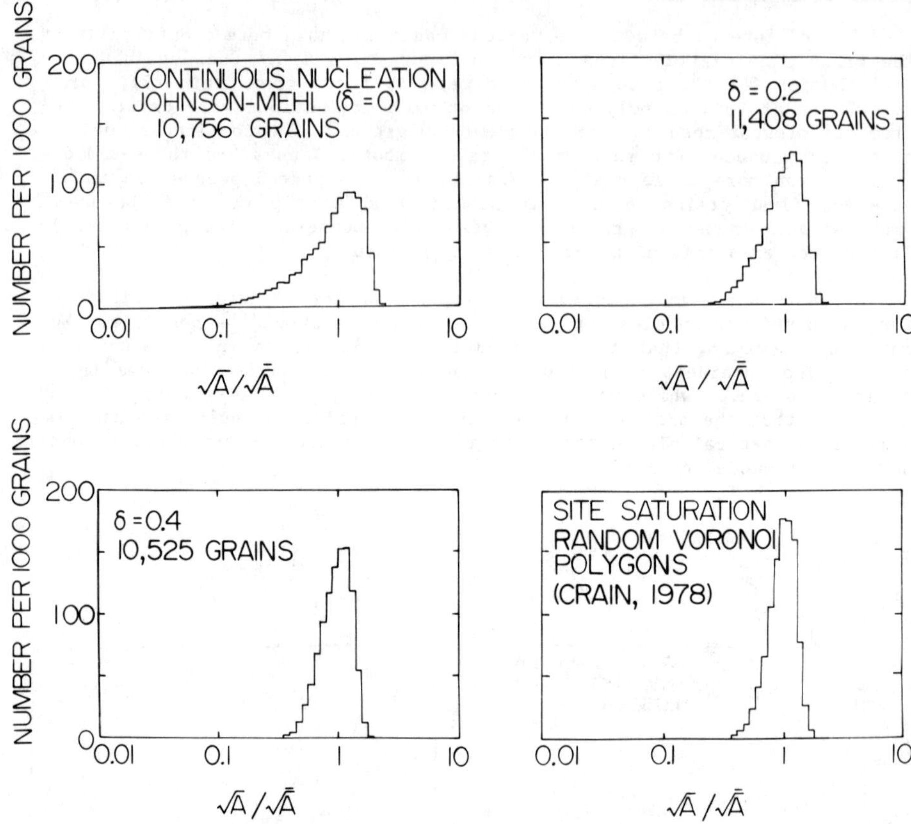

Figure 8: Distributions of grain size, as measured by the square root of the grain area, for the different nucleation conditions. These distributions are given on a logarithmic scale to facilitate the comparison to log normal distributions.

In Figure 8 we have replotted the grain size distributions on a logarithmic scale, using the square root of grain areas. This allows a direct comparison with the grain size distributions measured from thin films. The change from an area scale to a length scale changes the shape of the distributions, as does the change from linear to logarithmic.

Boundary Segment Lengths

Distributions of segment lengths, or boundary side lengths are shown in Figure 9. For the site saturation case, the distribution of segment lengths was reported by Crain (7). For the continuous nucleation case, the boundary segments are segments of hyperbolae. It is difficult to measure the actual length along the curve of a curved boundary in a micrograph. It is, however, straightforward to measure the total length of boundary per grain by measuring the number of boundary intercepts along random lines in

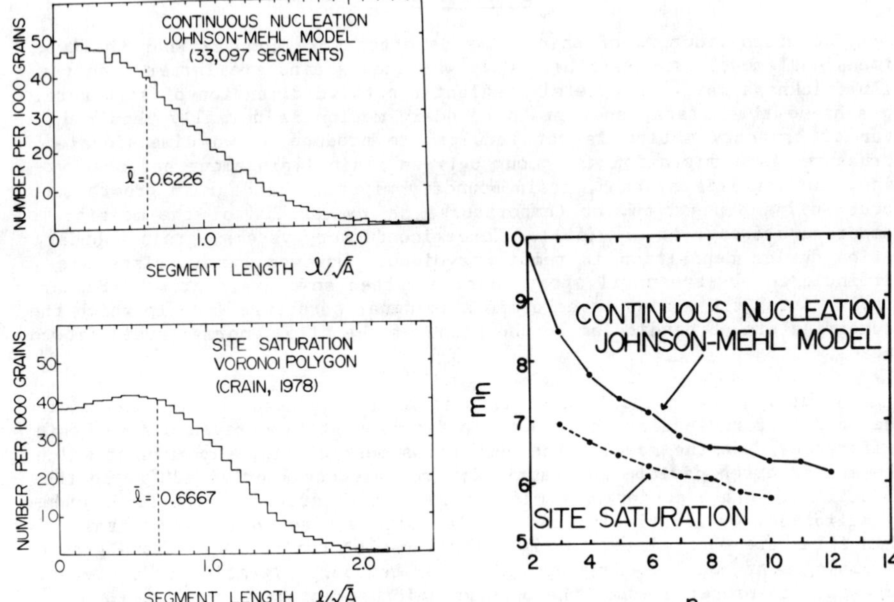

Figure 9: Distribution of segment length for the continuous nucleation and the site saturation structures.

Figure 10: The average number of sides, m_n, in the grains with different number of sides, n, for the site saturation or Voronoi polygon case (Boots, 1982), and the continuous nucleation case.

a micrograph with a known number of grains. Since there must be an average of six segments per grain, the average length of boundary per grain is directly related to the average segment length. We find that for the continuous nucleation case, the average segment length is 0.6226 (measured in units of the square root of average grain area). This agrees well with the theoretical prediction of 0.627 from Meijering (5). It is lower than for the random Voronoi polygons, which reflects the wider distribution in number of sides per grain. The average segment length also reflects the degree to which the grains are equiaxed. For the limiting case of an array of identical hexagons, the boundary segments are 0.6204 times the square root of hexagon area. For the depletion region models the average lengths are 0.6415 for $\delta = 0.2$ and 0.6480 for $\delta = 0.4$.

Number of Sides on Neighboring Grains

One topological parameter that has been measured for several cellular structures is the average number of sides on the grains that neighbor the grains with a particular number of sides. Recent studies of this include Lambert and Weaire ((8), (9)), Aboav ((10), (11), (12)), and Boots ((13)). Boots (14) calculated the distribution for the site saturation case. His results, along with those for the continuous nucleation case, are plotted in Figure 10. The dramatic difference is a reflection of the difference in the distributions of the number of sides per grain.

Discussion

The microstructure of thin films is often more complex than the two dimensional model can describe. Only when the grains are larger than the film thickness may we completely neglect the third dimension of structure. To achieve that state, some grain boundary motion is normally required. Even if boundary motion is not required to produce a two dimensional structure, some migration may occur between grain impingement and observation. In metallic systems, grain boundary migration and grain growth can occur during deposition, at temperatures as low as 20% of the melting temperature (Wong, et al. (15)). In semiconductor systems, grain boundary motion during deposition is readily avoided. However, grain sizes are difficult to observe until after there has been some grain growth (Palmer (16)). After the grains have formed a columnar structure (one in which the boundaries are perpendicular to the plane of the film), normal grain growth ceases.

A qualitative comparison of thin film microstructures with those of the models of nucleation and growth to impingement also reveals some basic differences. In the models, the boundaries meet at triple points at which the angles often deviate substantially from the equilibrium 120°. In the thin films, such a deviation would provide a high driving force for boundary migration. Qualitatively, the triple point angles in columnar grained thin films are much closer to 120° than those of our nucleation and growth to impingement models. Some provision for boundary migration is therefore necessary to properly model the observed microstructures.

There is another important reason for modelling the boundary migration. Most experiments report that normal grain growth in thin films essentially stops when the grain size becomes the film thickness. When the grains do grow larger than the film thickness, it is by the extensive growth of a few secondary grains, while the majority of grains remain unchanged. This process can be driven by differences in the energy of the free surface for different grains. It is clear than a very large grain, with a free surface energy lower than its neighbors, will grow by consuming its smaller neighbors, as described by Thompson (17). It is not exactly clear under what conditions a particular normal grain will become such a secondary grain. This can only be modelled by treating the details of local boundary migration.

We are developing a program which models grain boundary migration in two dimensions, driven by capillarity forces. It differs in technique from other recent models, such as the soap film model of Weaire and Kermode ((18), (19), (20)), and the grain growth model of Anderson, et al. (21) and Srolovitz, et al. (22). The Anderson, Srolovitz, et al. model is based on a large array of incremental areas, each of which belongs to one grain or another. In that model, the structure evolves as each incremental area is repeatedly allowed to change from one grain to another, based on random chance, weighted according to the grain membership of the neighboring incremental areas. Although their model results in grain growth, with boundary migration rates which are, on average, proportional to curvature, it is not accurate on a very local scale.

We have, therefore, chosen to follow the boundary itself as a linear array of moving points. In each time increment, the points are moved, perpendicular to the boundary line, by an amount proportional to the local curvature. The program provides for increasing or decreasing the number of points as required to keep the spacing between points along the line within certain bounds. It also provides for the elimination of small grains and the switching of grain neighbors. The program alternates between moving

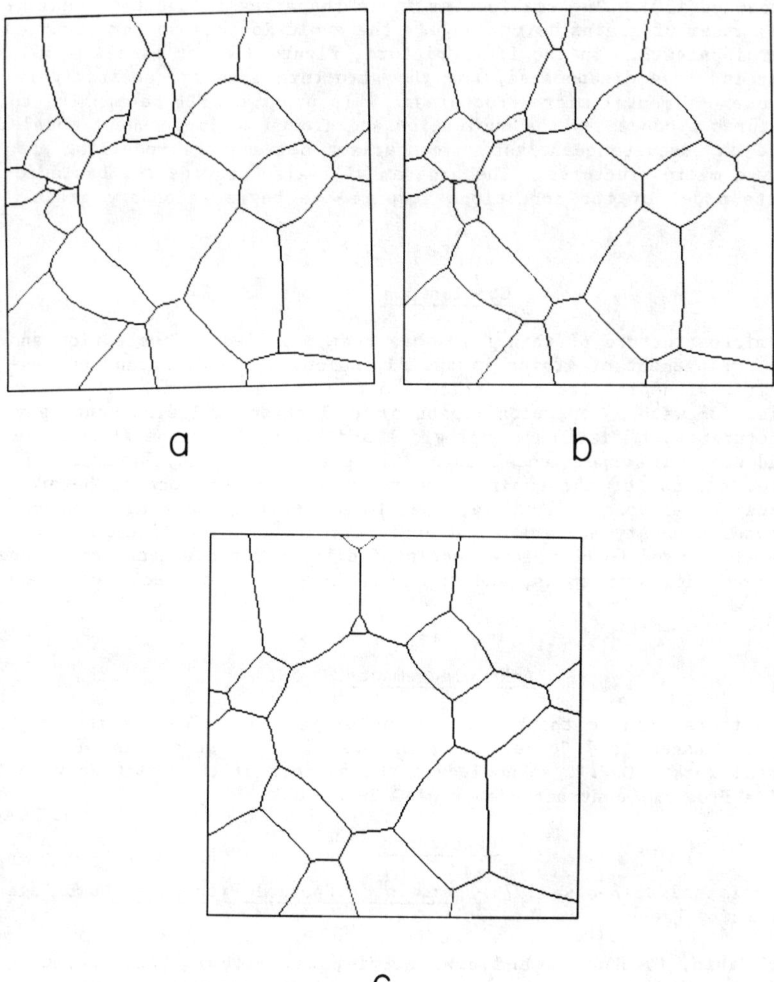

Figure 11: Demonstration of the program which models grain boudary migration. (a) Initial structure, chosen arbitrarily. (b) Structure after 50 time increments of boundary migration. (c) Structure after 500 time increments of boundary migration.

the boundary segments, and moving the triple points to positions which maintain local angles of 120°.

We here present the preliminary results on the evolution of an arbitrary initial array in which the boundary curvatures and triple point angles deviate markedly from local equilibrium. Figure 11a shows the starting configuration. The array is 750 x 750 length units, and the boundary mobility is taken as 12 length units squared per time unit. In this model, the boundaries are fixed at the points where they meet the margins of the model. At time t = 50 (Figure 11b), the array has evolved so that the boundaries are noticeably smoothed, and the triple point angles have approached 120°. The smallest grain in the array is about to vanish. Since the number of grains is unchanged, the evolution has so far occurred without grain growth. In the final picture, Figure 11c, at time t = 500, several grains have disappeared, and the structure appears qualitatively similar to experimental microstructures. This program will be applied to the structures produced by the nucleation and growth to impingement models, to provide an adequate quantitative comparison between the modelling and the observed microstructures. The program will also provide the basis for an accurate model of the conditions required to begin secondary grain growth.

Conclusions

The microstructure of thin films has been modelled by nucleation and growth to impingement of grains in two dimensions. The different nucleation conditions used: site saturation, continuous nucleation, and continuous nucleation with a depletion region or nucleation exclusion zone, give microstructures with distinct topological and geometric properties. The correspondence with experimental thin film microstructures, however, is inadequate, because of the grain boundary migration that occurs before experimental observation. We have, therefore, created the program necessary to model boundary migration and grain growth on a local scale. This program will be used to correlate models of different nucleation conditions with observed microstructures, and to model the process of secondary grain growth.

Acknowledgements

The authors wish to thank Junho Whang of Dartmouth College for programming assistance, and Joyce Palmer of M.I.T. for discussions of her experimental work. C.V.T. acknowledges the support of the Joint Services Electronics Program, contract number DAAG-29-83-K003.

References

1. A. Getis and B. Boots (1979), Models of Spatial Processes, Cambridge University Press

2. K.W. Mahin, K. Hanson, and J.W. Morris, Jr. (1980), "Comparative Analysis of the Cellular and Johnson-Mehl Microstructures through Computer Simulation", Acta Met., 28, 443-453

3. W.A. Johnson and R.F. Mehl (1939), "Reaction Kinetics in Processes of Nucleation and Growth", Trans. AIME, 135, 416-458

4. U.R. Evans (1945), "The Laws of Expanding Circles and Spheres in Relation to the Lateral Growth of Surface Films and the Grain-Size of Metals", Trans. Faraday Society, 41, 365-374

5. J.L. Meijering (1953), "Interface Area, Edge Length, and Number of Vertices in Crystal Aggregates with Random Nucleation", Philips Res. Rep., 8, 270-290

6. E.N. Gilbert (1962), "Random Subdivisions of Space into Crystals", Annals of Mathematical Statistics, 33, 958-972

7. I.K. Crain (1978), "The Monte-Carlo Generation of Random Polygons", Computers and Geosciences 4, 131-141

8. C.J. Lambert and D.L. Weaire (1981), "Theory of the Arrangement of Cells in a Network", Metallography, 14, 307-318

9. C.J. Lambert and D.L. Weaire (1983), "Order and Disorder in Two-Dimensional Random Networks", Phil. Mag. B, 47, 445-450

10. D.A. Aboav (1980), "The Arrangement of Cells in a Net", Metallography, 13,43-58

11. D.A. Aboav (1984), "The Arrangement of Cells in a Net. III", Metallography, 17, 383-396

12. D.A. Aboav (1985), "The Arrangement of Cells in a Net. IV", Metallography, 18, 129-147

13. B.N. Boots (1985), "Further Comments on "Aboav's Rule" for the Arrangement of Cells in a Network", Metallography, 18, 301-303

14. B.N. Boots (1982), "The Arrangement of Cells in 'Random' Networks", Metallography, 15, 53-62

15. C.C. Wong, H.I. Smith, and C.V. Thompson (1985), Presented at the Spring Meeting of the Materials Research Society, San Fransico

16. J.E. Palmer (1985), "Secondary Grain Growth in Ultra Thin Germanium Films on Silicon Dioxide", M.S. Thesis, M.I.T.

17. C.V. Thompson (1985), "Secondary grain growth in thin films of of semiconductors: Theoretical aspects", J. Appl. Phys., 58, 763-772

18. D. Weaire and J.P. Kermode (1983a), "Computer Simulation of a Two-Dimension Soap Froth, I. Method and Motivation", Phil. Mag. B, 43, 245-259

19. D. Weaire and J.P. Kermode (1983b), "The Evolution of the Structure of a Two-Dimensional Soap Froth", Phil. Mag. B, 47, L29-L31

20. D. Weaire and N. Rivier (1984), "Soaps, Cells and Statistics -- Random Patterns in Two Dimensions", Contemporary Physics, 25, 59-99

21. M.P. Anderson, D.J. Srolovitz, G.S. Grest, and P.S. Sahni (1984), "Computer Simulation of Grain Growth--I. Kinetics", Acta Met., 32, 783-791

22. D.J. Srolovitz, M.P. Anderson, P.S. Sahni, and G.S. Grest (1984), "Computer Simulation of Grain Growth--II. Grain Size Distribution, Topology, and Local Dynamics", Acta Met., 32, 793-802

COMPUTER MODELS OF FROTHS AND VORONOI PATTERNS

D Weaire and J Wejchert

Department of Pure and Applied Physics
Trinity College
Dublin 2

Abstract

We present some of the recent findings in our study of two-dimensional cellular systems, which are related to microstructures and their evolution. We are mainly concerned with the 2d soap froth system. With the aid of computer simulations, in which Voronoi networks play a useful role, we study the processes involved in the evolution of the soap froth and make statistical inferences from our simulation, particularly regarding asymptotic behaviour. The soap froth has often been regarded as a useful analogue of grain structure in metals. However, the basic equations which govern its evolution are not exactly the same. Hence, the asymptotic behaviour need not be identical in the two systems. More experimental and theoretical work is needed before definitive statements can be made about either case or their comparison.

Introduction

Microstructures, in general, pose intriguing problems of geometry and topology related to their random nature. These may be difficult to analyse even when the basic underlying mechanisms of formation of the observed patterns are known. If one studies the grain structure of metals and polycystalline materials, or domains in glassy substances, one notices a variety of configurations. One observes that the structures divide space into cells (or domains) with diversity of cell area, shape and local topology. Nevertheless, different systems often exhibit remarkably similar underlying patterns. This may be attributed to the operation of common physical principles, particularly minimisation of boundary energy. Also, most observed microstructures are the result of some evolutionary process such as annealing, which brings the structure to an asymptotic form largely independent of its starting configuration.

This was the reasoning which led Smith (1) to suggest the 2d soap froth system as a prototype for a wide class of such structures. We have constructed computer models of this system, originally with a view to determining the asymptotic behaviour. With computer simulation we can study the evolution of the structure with time, which gives us insight into the temporal changes of microstructures in general (2), (3).

The major quantities of interest are of a statistical nature: the frequency distributions of cell areas and numbers of cell sides, and their respective moments about the mean. It may be important to establish, for example, whether the system becomes stable in the topological sense during the course of time, i.e. whether the distributions of topological quantities reach stable forms for $t \to \infty$, or whether the network becomes progressively more and more disordered.

The experimental results of Smith for the 2d soap froth suggested that the asymptotic behaviour (i.e. $t \to \infty$) is stable, so that parameters such as the second moment ($\mu_2{}^s$) of the distributions of sides f(n) remained constant. Furthermore, it was found that the mean area of cells varied linearly with time ($A \sim t$ or the average diameter $d \sim t^{1/2}$). However, a later and apparently preferable analysis by Aboav (4), which was based on Smith's photographs, was at variance with the above conclusions. Aboav's analysis suggested that $\mu_2{}^s$ increases linearly with time and that $A \sim t^2$.

By using computer models we can take our level of abstraction one step further and rid ourselves of possible extraneous factors in experiment. Our model soap froth is defined as follows:

a) Two-dimensional space is divided into cells of fixed area, since they contain an incompressible gas.

b) The total boundary length is minimised at all times (corresponding to the minimisation of surface energy). The process by which this takes place in experiment or simulation will here be called "equilibration".

c) The temporal evolution of the system takes place due to the diffusion of gas between cells (according to Von Neumann's law - see later), i.e. the constraint of prescribed cell areas changes with time.

d) Periodic boundary conditions are used in conjunction with a finite sample, in order to represent as closely as possible the properties of an inifinite system.

A consequence of a) and b) (see Weaire and Rivier (5)) is that the vertices in our network are three-fold and that the equilibrium angle between any vertex edge is $2\pi/3$. Figure (1) shows a typical configuration of soap froth cells from our simulation.

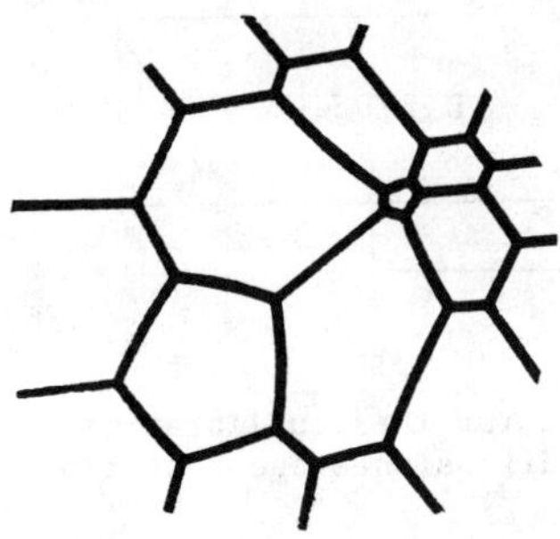

Figure 1 - Typical configuration of soap cells in equilibrium. Notice that any vertex angle is $2\pi/3$.

Diffusion is invoked through Von Neumann's law (6), which is solely dependent on local topological properties. This states that

$$dA/dt = K(n-6)$$

where A is the area of a cell, n is its number of sides and K is a constant. Note that for the network to change according to the above, we must have some cells for which $n \neq 6$. Subsequently, cells with $n < 6$ will shrink and eventually disappear and cells with $n > 6$ will grow and gain sides so that the structure is continually changing.

In practice, by virtue of the fact that equilibration can be taken to be instantaneous with respect to diffusion, the two distinct procedures are carried out separately in a repeating cyclic fashion. Figure (2) shows a simple flowchart for the above process.

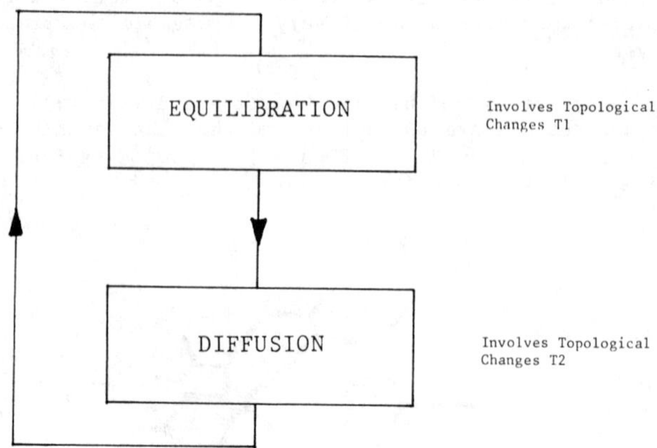

Figure 2 - Flowchart for the equilibration - diffusion cycle. It is during the diffusion stage that time is incremented.

Equilibration and evolution in accordance with the above principles entail topological changes which can cause difficulties in programming. These are the T1 and T2 processes (in the nomenclature of Weaire and Rivier (5)). In these events topological rearrangements in the network change the local environment of adjacent cells (T1), or a cell shrinks and disappears completely (T2). Figure (3) illustrates these processes.

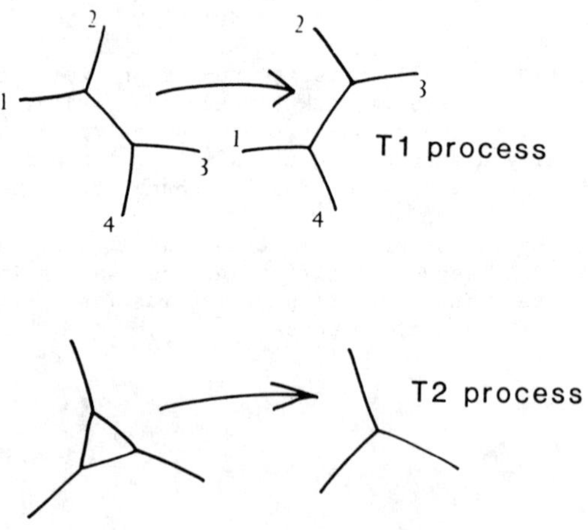

Figure 3 - The T1 and T2 processes. The first is a local rearrangement of cells, the second, constitutes the vanishing of a cell.

The Soap Froth Models

In the course of our investigations, two quite different computer programs were written to simulate the froth. The first and primary simulation modelled the system by concentrating on the vertices, relative angles and areas of the cells *per se*; this method we shall subsequently call the continuous simulation. The second program represented the network on a discrete lattice and used a Monte Carlo algorithm for equilibration.

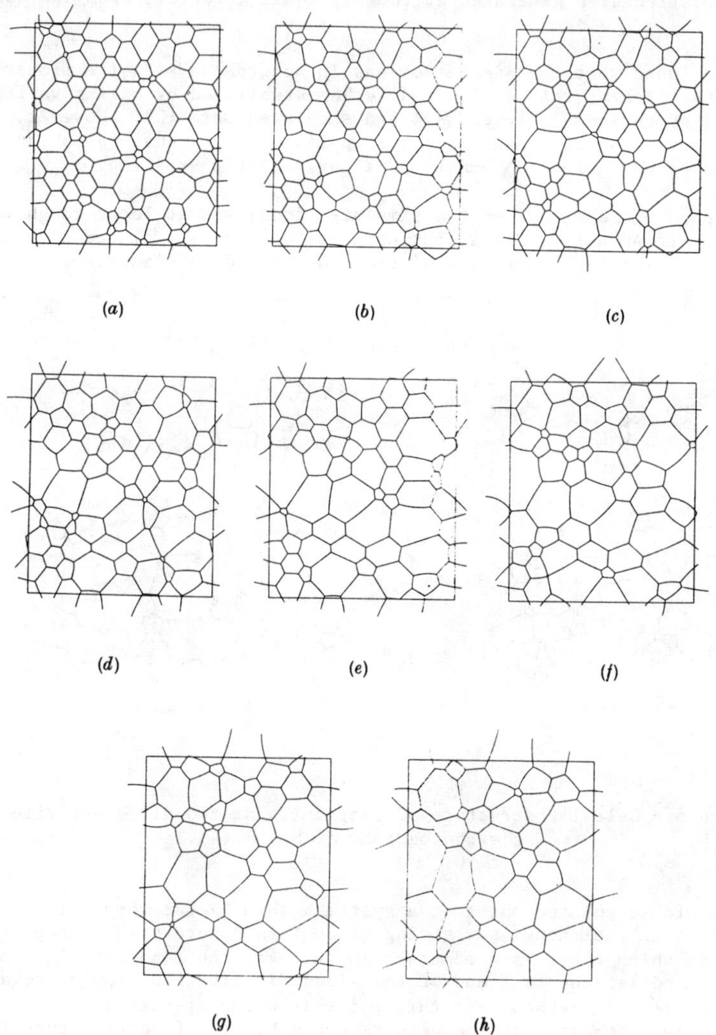

Figure 4 - Sequence of soap froth cell structures at roughly equal times.

The continuous program was able to illustrate most of the salient features of the system. Its method was to relax the local environment of a vertex towards equilibrium, for each vertex in turn, until an overall equilibrated configuration was reached. Subsequently, increments of area were transferred between cells, simulating diffusion as prescribed by Von Neumann's law. The equilibration - diffusion cycle was then continuously repeated, thus advancing the evolution of the network. Figure (4) shows a sequence of structures generated at roughly equal intervals from the above model.

It was found that the simulation was in accordance with the properties of a real soap froth; this could readily be established by making various statistical comparisons between real and simulated data (2). However, apart from such a purely quantitative comparison, the visual resemblance of the two (compare figures (4) and (5)) is particularly striking.

The temporal behaviour of the simulated froth was at least roughly consistent with Aboav's results, but computational limits placed a severe restriction on the maximum period of time that could be simulated.

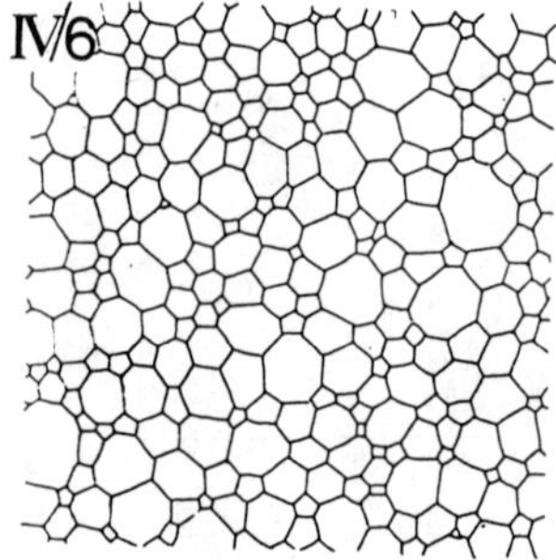

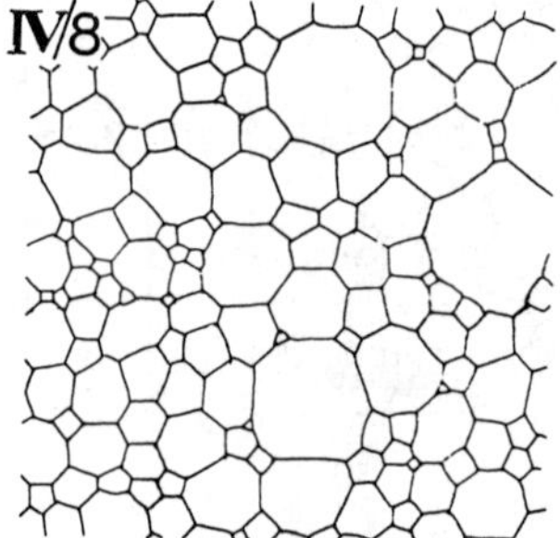

Figure 5 - Cellular structure of soap cells at two different times from the photographs of Smith (4).

It should be pointed out that a system with a linear temporal dependence of μ_2^s, although surprising, is not an absurdity. It represents a network in which there is a constant increase in the topological diversity of cells; the tail end of the sides distribution growing towards higher n, as time increases. At this point it was suggested that a possible asymptotic form for the network could be of a fractal nature (5). Such a system could indeed have an arbitrarily high μ_2^s; see Figure (6). This provides a further motivation for the study of soap froth.

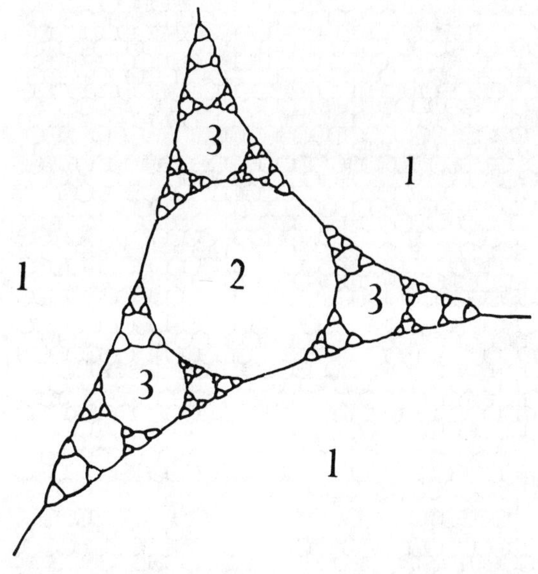

Figure 6 - Illustration of a fractal structure for a soap froth system.

In an effort to throw further light on the asymptotic behaviour, a second approach to simulation was undertaken, employing the Monte Carlo method (7). In this case we represent the network of cells on a discrete lattice of points, each of which has six nearest neighbours. Clustered groups of lattice points labelled with integers represent the cells. Changes in cell geometry and topology are expressed as the relabelling of points and so T1 and T2 processes are implicitly included, and pose no programming problems. The set of lattice points can be stored and manipulated as a two-dimensional array in the computer program. Figure (7) shows the network mapped on to the array of lattice points. By extending the basic principle of surface energy minimisation to include the constraint of constant cell area, we equilibrate the froth network. Cells are defined to have specific target areas prior to equilibration and deviations from targets entail changes in energy. Therefore, the total energy of the system is represented by both surface and area terms. Lattice points are randomly switched between their nearest neighbours, only if the difference in associated energy is unchanged or reduced. This corresponds to carrying out the standard Monte Carlo rule at zero temperature.

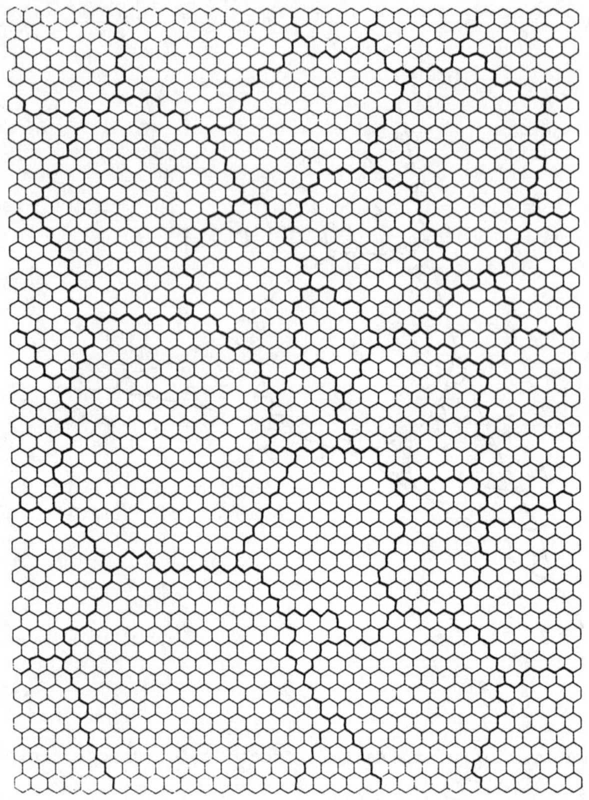

Figure 7 - Network of cells represented on a discrete lattice of points (the centres of the hexagonal cells).

As a result, slight changes in cell areas and configuration take place continually until an equilibrated configuration is achieved. Once again Von Neumann's law is used to simulate the growth process. In this case the target areas of the cells are simply reset to the new values. Thereafter, in similar spirit to that of the continuous simulation, the system evolves by looping around the equilibration and diffusion processes.

The extra term in the total energy may be regarded as introducing a finite compressibility of the gas in the cells; the magnitude of this requires to be chosen. At one extreme we have the situation where cells are completely incompressible, and the allowed changes are severely restricted, resulting in poor convergence of the method. At the other there is no constraint of constant cell area. A practical compromise lies between these two extremes. In principle the compressibility should approach zero; however, in this case, exorbitant amounts of central processing time have to be used to bring the system to an equilibrium configuration. At the other extreme, of the extra energy term being set equal to zero, there is no constraint of constant cell area. Our results for this case, are in complete agreement with the grain growth model of Srolovitz et al (8).

In general, the early behaviour of the Monte Carlo method was found to be in keeping with the continuous simulation. However, for times greater than those the previous simulation could model, the resulting behaviour

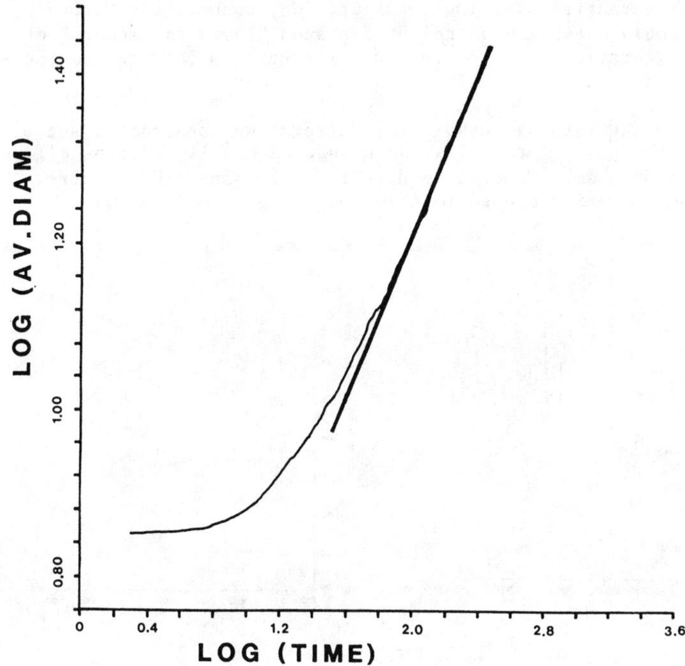

Figure 8 - Log plot of average cell diameter against time. The line drawn indicates the linear region for longer time behaviour.

differed significantly. In fact it supported Smith's original estimate, namely that $A \sim t$ and that $\mu_2{}^s \sim$ constant! Figure (8) shows how, by taking log plots of average area v time, we found an accurate exponent in the growth law $d = kt^c$; c was found to be .49 ± .05, suggesting that $d \sim t^{1/2}$.

The observation that $\mu_2{}^s$ does not continue rising indefinitely indicates that the distribution of sides tends to a stable form for $t \to \infty$.

Thus, it can be seen that the asymptotic behaviour of this stimulation can be described (on average) as a change in scale of the system without change in topological distributions.

Clearly one may further ask: what are the starting configurations of these simulations of random structures? Is there any relationship between a particular starting configuration and the ensuing evolution?

In the case of the continuous simulation we initially chose a perfect hexagonal network as our starting point. This had to be perturbed slightly to introduce some topological disorder so that diffusion could take place. By using an alternative starting configuration, such as the Voronoi network, we could establish whether the system "loses the memory" of its original configuration i.e. do the two alternatives converge to the same behaviour at later times, and if so, how long does this take?

To create the Voronoi network (or pattern) we construct a set of lines which divide the plane into cells, such that each cell contains all points closer to one of a set of predefined centres than any other centre. Details of our algorithms used to this end can be found in (3).

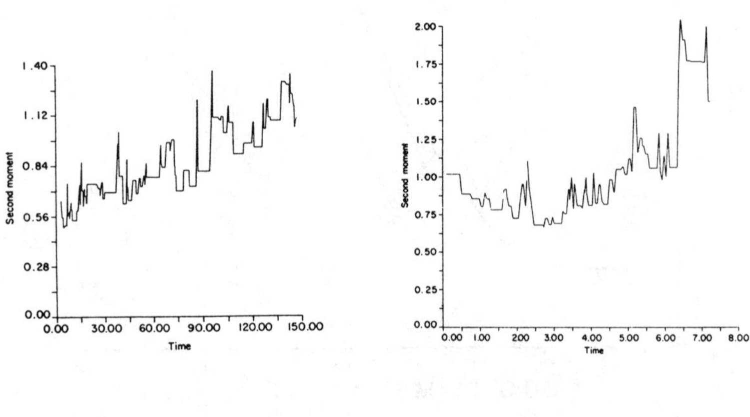

(a) (b)

Figure 9 - The second moment of the distribution of sides, $\mu_2{}^s$ of the network for: (a) the hexagonal and (b) the Voronoi starting, configurations.
The time values in (b) should be multiplied by 20 for consistency.

The two plots in Figure (9), show the second moment $\mu_2{}^s$ traced throughout two simulations, each with different starting networks. As it can be seen, both appear to exhibit the same (linear) behaviour after sufficient time has elapsed.

Voronoi Networks

Since the Voronoi network has been popular for a variety of reasons among materials scientists, it may be helpful to add some comments regarding its specific properties. Some of the characterising parameters of the network are known exactly; however, others have not yielded to a purely quantitative explanation. As example of such an exact result, Gilbert (9) has shown that the second moment of the distribution of areas is given by

$$\mu_2 A = 0.28 A_m^2$$

where A_m is the mean cell area.

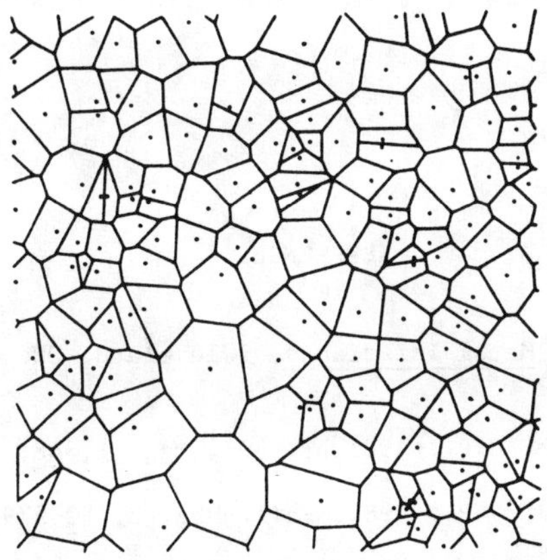

Figure 10 - An example of a Voronoi pattern; the dots mark the centres of cells.

However, no exact result is known for the area distribution itself. Kiang (10), found that the gamma distribution gave a good fit to numerically generated data of cell areas. In a more recent effort (11), we presented a rough argument, that leads to the above theoretical form. Our investigation was backed by computer generated Voronoi networks, of both the continuous and discrete methods. We confirmed that the gamma distribution does give an excellent fit to generated data, and we also found its governing exponent with high precision.

The starting point for our derivation was the probability distribution for a Poisson point process; this is a distribution function for distances between randomly scattered points in the plane. The distribution was then self-convoluted and subsequently converted into a distribution of areas. Our above argument is not rigorous; nevertheless, we have gained further insight into the remarkable agreement that is found between the actual area distribution and the suggested theoretical form.

Conclusion

Originally, one of our principle aims was to understand the aymptotic behaviour of the soap froth. Despite the success of the continuous method in simulating the salient features of this system, it proved inadequate to model the network for $t \to \infty$, with our present computer resources.

As might have been expected, our subsequent Monte Carlo approach proved more successful in this respect. However, the simulation method itself still requires attention, in particular regarding the degree of cell compressibility allowed. Therefore, in a sense, our computational investigations are still inconclusive and require further study. Moreover, more experimental investigation has to be undertaken, since we remain faced with rather old and contradictory data.

REFERENCES

1. Smith, C.S., 1952, Metal Interfaces, (Cleveland, OH: American Society for Metals), p. 65

2. Weaire, D., Kermode, J.P., 1983 b, Phil. Mag. B, 48, 245

3. Weaire, D., Kermode, J.P., 1984, Phil. Mag. B, 50 379

4. Aboav, D.A., Metallography, 13, 43

5. Weaire, D., Rivier, N., 1984, Contemp. Phys., 25, 59

6. Von Neumann, J., 1952, Metal Interfaces, (Cleveland, OH: American Society for Metals), p. 108

7. Wejchert, J., Weaire, D., Kermode, J.P., (to be published, Phil. Mag.)

8. Srolovitz, D.J., Anderson, M.P., Sahni, P.S., Grest, G.S., 1984, Acta Metall. 32, 793

9. Gilbert, E.N., 1962, Ann. Math. Stat. 33, 958

10. Kiang, T., 1966, Z. Astrophys., 64, 433

11. Weaire, D., Kermode, J.P., Wejchert, J., (to be submitted, Phil. Mag.)

COMPUTER SIMULATED STAGNATION AND ABNORMAL GRAIN GROWTH

G. Abbruzzese

Centro Sperimentale Metallurgico S.p.A., C.P. 10747 00100 ROMA-EUR (Italy)

Abstract

A statistical model, to describe grain growth in metals and alloys has been developed. With this model, by investigating the evolution of the grain size distribution, it is possible to follow the growth in the presence of second phase particles. Under strong growth inhibition (Zener drag), the conditions for complete growth stagnation are achieved. In this case $I_R R_c > 1$ (I_R = intensity of the inhibition, R_c = critical radius of the size distribution) and the main features of the grain size distribution are the following: the smallest grains, up to a value which is a reciprocal function of inhibition, disappear; the stronger the inhibition, I_R, leading to stagnation the more the shape of the initial distribution affects the final one; the size distribution develops a maximum corresponding to the smallest grains size (the exact position depends on the initial distribution); when $I_R R_c \longrightarrow 1$ the shape of the distribution appears to be independent on the initial distribution.

When $2/3 < I_R R_c < 1$ the computer simulation suggests the presence of abnormal growth. In this way the results of the bidimensional Hillert model, founded on topological considerations, are confirmed.

Introduction

Grain growth in metals and alloys can be of two kinds, namely: normal and abnormal grain growth or following a more recent terminology continuous and discontinuous grain growth.

In the past, normal grain growth has been extensively treated (1-4). Fundamental results of these works are the existence of a quasi-stationary distribution and the $\bar{R} = k\ t^{1/2}$ law for the growth of the average grain. Some differences in the shape of the stationary distribution have been found depending on the driving-force adopted for grain growth.

Abnormal grain growth has been related to the presence of second phase particles which act as a restraining force (Zener drag) to the growth (*)(1,6-8). In this case critical conditions can be set up in which some grains, greater in size than the average, have the capability of growing at the expense of the matrix grains. After an incubation time necessary for the growing grains to achieve an adequate size and to develop a heterogeneous structure, the process develops with an acceleration of growth until the establishment of a new condition of substantial structural homogeneity (large grains with low size dispersion). When the restraining force is sufficiently strong the grains, after a certain amount of growth, can achieve an equilibrium condition (grain growth stagnation). In that case the driving force acting on the growing (or shrinking) grains is equivalent to the frictional force produced by the second phase particles dispersed in the matrix (9).

Recently it has been shown that it is possible to treat grain growth by a statistical model taking into account the effect of the Zener drag (9). Moreover some analytical conditions which characterize the grain growth stagnation were determined. The basic hypotheses of the model are: i) the material is considered to be made up of spherical grains and the driving force for grain growth is the reduction of the grain boundary area and it acts isotropically with a pressure equal to $2\gamma/R$ on the boundary of each grain (γ is the specific boundary energy and R the grain radius). The pressure on the boundary between two grains is simply given by the difference: $2\gamma/R_1 - 2\gamma/R_2$. ii) a given grain size distribution is divided in size classes and each of them contains equal sized grains. iii) each grain belonging to j-th class has the same surrounding grains therefore they behave in a similar way (homogeneity hypothesis which is essential to a statistical model). iv) each grain belonging to the j-th class has a surrounding of randomly distributed grains (it is possible to show that this hypothesis is not necessary), therefore the contact area between a grain of the j-th class and all grains of

(*) The discontinous growth can be also controlled by the texture present in the material(5), but in this paper only a matrix with random oriented grains will be considered and the mechanisms of the secondary recrystallization induced by particles will be pointed out.

the j-th class is given by:

$$a_{ij} = \frac{n_j R_j^2}{\sum_{j=1}^{n_c} n_j . R_j^2} . 4 \pi R_i^2 \qquad i,j = 1,2...,n_c \quad (1)$$

where n_i, n_j are the number of grains in the i-th and j-th size classes, respectively, and n_c is the number of the size classes considered.

The equation giving the rate of growth of grains in size class i is obtained by a volume conservation relationship valid for two grains in contact and averaged over all neighbouring grains by the contact areas a_{ij}:

$$\frac{dR_i}{dt} = - \frac{M}{4 \pi R_i^2} \sum_{j=1}^{n_c} a_{ij} \left(\frac{1}{R_i} - \frac{1}{R_j}\right) \quad i = 1, 2, ..., n_c \qquad (2)$$

where $M = 2 m \gamma$ (cm^2/sec) and m is the boundary mobility.

Moreover the movement of the grains in the (R, t) space is controlled by the continuity equation:

$$\frac{\partial f}{\partial t} + \frac{\partial}{\partial R} \left(f \frac{dR}{dt}\right) = 0 \qquad (3)$$

written in a discrete form (9).

Equation (3) expresses nothing but the local grain number conservation in (R, t) space; that is, the variation of the number of grains of size R due to the flux of the incoming and outgoing grains from and toward the contiguous size classes in a certain time interval Δt. The possibility of using this flux equation in the (R, t) space for the movement of the grains is the basis of the statistical model and is allowed by the homogeneity hypothesis (each grain of size class R has the same surrounding, therefore all the grains of size R behave dynamically in the same way).

Solving equation (2) numerically together with equation (3) allows us to follow the evolution of the grain size distribution as a function of the time. Adding the terms inside the brackets of equation (2) yields:

$$\frac{dR_i}{dt} = M\left(\frac{1}{4\pi R_i^2}\sum_{j=1}^{n_c}\frac{a_{ij}}{R_j} - \frac{1}{R_i}\right) \qquad i = 1, 2, \ldots\ldots\ldots, n_c \tag{4}$$

where the quantity

$$R_c = \frac{4\pi R_i^2}{\sum_{j=1}^{n_c}\frac{a_{ij}}{R_j}} \tag{5}$$

has the same meaning of critical radius as in (1). When an inhibition of grain growth I_R is present equations (4) changes as follows:

$$\frac{dR_i}{dt} = M\left(\frac{1}{R_c} - \frac{1}{R_i} \pm \frac{I_R}{2}\right) \qquad i = 1, 2, \ldots\ldots\ldots, n_c \tag{6}$$

the sign + for grains whose R_i satisfies: $1/R_c - 1/R_i < 0$

the sign - for grains whose R_i satisfies: $1/R_c - 1/R_i > 0$

and $\frac{dR_i}{dt} = 0$ when

$$\left|\frac{1}{R_c} - \frac{1}{R_i}\right| \leqslant \frac{I_R}{2} \tag{7}$$

On substituting (1) in equation (5) we get:

$$R_c = \frac{\sum_{j=1}^{n_c} n_j R_j^2}{\sum_{j=1}^{n_c} n_j R_j} = \frac{\overline{R^2}}{\bar{R}} = \frac{\bar{R}^2 + \sigma^2}{\bar{R}} \tag{8}$$

where $\bar{R}$ and σ^2 are the average radius and the variance of the grain size distribution, respectively.

It has been shown that in the presence of strong inhibition, I_R, grain growth stops ; the validity of equation (7) is extended to all grains of the distribution (grain growth stagnation) and the following independent equations describing this equilibrium condition may be written:

$$R_c = 2\,\frac{R_{MAX} \cdot R_{MIN}}{R_{MAX} + R_{MIN}} \tag{9}$$

$$I_R = \frac{1}{R_{MIN}} - \frac{1}{R_{MAX}} \quad \text{or} \quad I_R = 2\left(\frac{1}{R_c} - \frac{1}{R_{MAX}}\right) \tag{10)a,b}$$

To verify the validity of equations (8), (9) and (10a,b) the grain size distribution has been measured on Fe-3%Si material, out of equilibrium (before the stagnation is set up) and after equilibrium is achieved. The inhibition factor has been evaluated by measuring the volume fraction of the second phase by chemical analysis and the size of the particles by electron microscopy. The agreement between the theoretical grain distribution evolution and the experimental one is quite good and the validity of equations (9) and (10) are confirmed (9).

In this paper the meaning of the equations (9) and (10) will be examined in more detail in order to determine the conditions for abnormal grain growth (secondary recrystallization) in a homogeneus matrix of randomly oriented grains. Moreover the meaning of these criteria will be analyzed in comparison with the results coming out from some numerical simulations.

Computer simulations of grain growth in presence of second phase particles

The curve in Fig. 1 is adopted as an initial grain size distribution ($t = 0$) (with $\bar{R} = 5.3\ \mu m$ and $R_{MAX}/\bar{R} = 2.9$) to carry out some computer simulations with different levels of restraining force (inhibition I_R in the range 0-1600 cm^{-1}). The simulations were carried out with a rate coefficient $M = 2\, m\gamma = 4\,\pi\, 10^{-9}\ cm^2/sec$. In Fig. 2 the mean grain radius vs time, for different level of inhibition I_R, and in Fig. 3, 4 the evolution of the grain size distribution, are shown. The mean radius is continuously growing, for $0\ cm^{-1} \leq I_R \leq 400\ cm^{-1}$, with a growth law which, as the inhibition increases, moves away from the $\sim t^{1/2}$ law valid for $I_R = 0$ (the exponent becomes lower than .5).

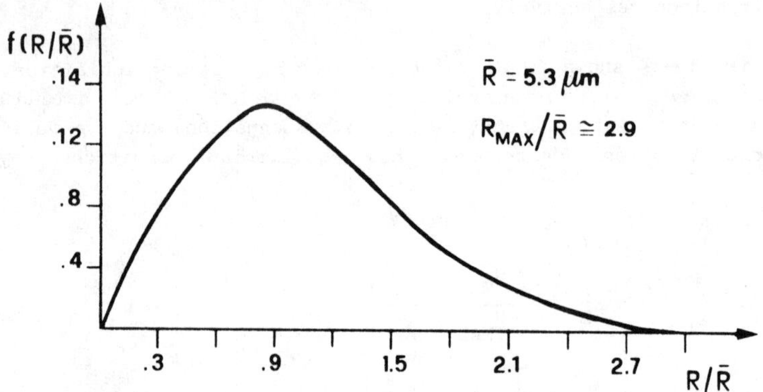

Figure 1 - Initial size distribution (t = 0) adopted in the computer simulations

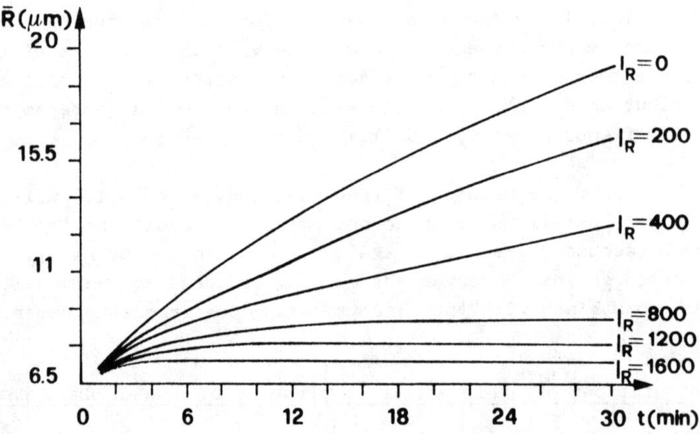

Figure 2 - mean radius vs time for different degrees of inhibition I_R (cm^{-1}). The value $M = 4\pi\ 10^{-9}$ (cm^2/sec) is adopted for the rate coefficient and the simulation time up to 30 minutes is shown

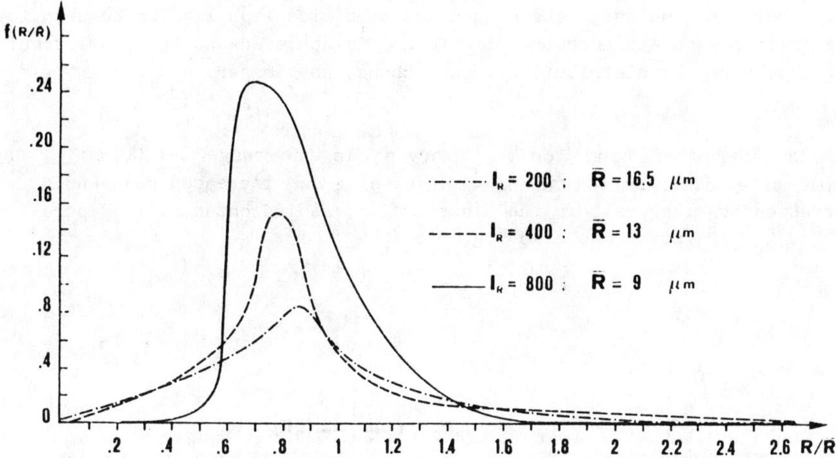

Figure 3 - Grain size distributions after 30' of simulation time for different amounts of inhibition (I_R = 200 - 800 cm^{-1})

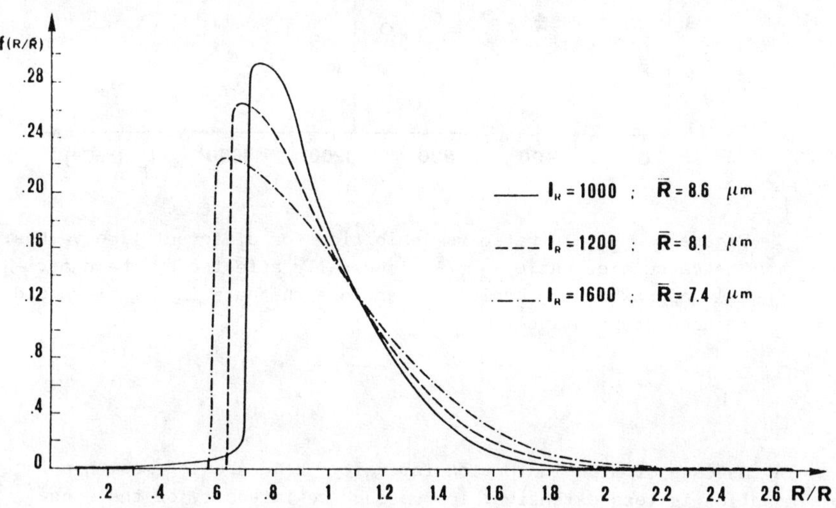

Figure 4 - Grain size distributions after 30' of simulation time for different amounts of inhibition (I_R = 1000 - 1600 cm^{-1}). For I_R = 800 - 1000 cm^{-1} the distributions do not correspond to complete achievement of grain growth stagnation (see also Fig. 2)

Moreover when I_R is greater than $\sim 1000\ cm^{-1}$ grain growth proceeds for a certain time until the stagnation condition (equilibrium between total grain pressure and restraining force) is achieved and the statistical parameters of the distribution do not change any longer.

On the other hand for I_R varying in the range 0-1000 cm^{-1} the grain size distribution changes; developing an increased peak and a corresponding long tail in the distribution, as indicated by the increase in the $R_{MAX}/\bar{R}$ ratio (see also Fig. 5).

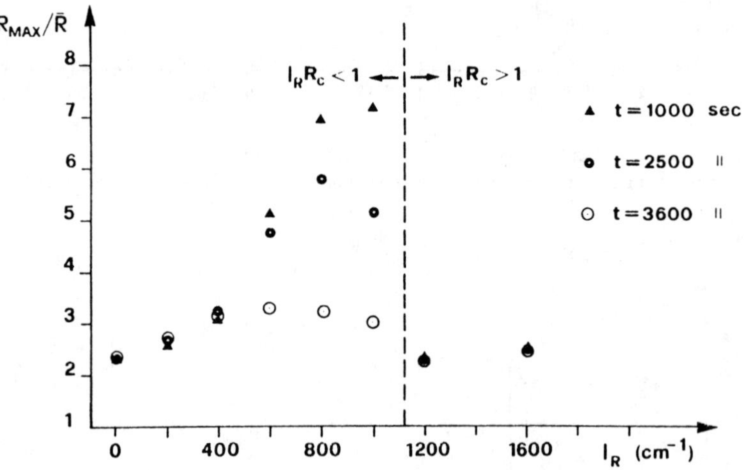

Figure 5 - $R_{MAX}/\bar{R}$ ratio vs inhibition for different time values. The calculated ratio $R_{MAX}/\bar{R}$ is usually affected by the numerical approximation achieved therefore the data in figure should be used comparatively

In the case of high inhibition intensity the degree of growth prior to stagnation is less extensive. Increasing the intensity of the Zener drag results in a change in the size distribution: with increasing inhibition the peak height decreases, the position of the maximum is shifted toward lower $R/\bar{R}$ values (less small grains consumed), and the ratio $R_{MAX}/\bar{R}$ decreases (see Fig. 4)

In fact under stagnation conditions, the final grain size distribution is in general affected by the initial distribution more and more with increasing inhibition because a strong I_R value leads to the disappearence of few smaller grains and hence to a limited increase in the volume of bigger ones. Consequently the shape of the initial distribution corresponding to the bigger grains is not significantly modified.

The increase in size of the bigger grains correspond to consumption of a large number of smaller grains (the volume of the system remain constant); this leads to a smaller distribution width (a smaller variance and $R_{MAX}/\bar{R}$ ratio; see Fig. 5 and 6) and a higher weight in the system of the grains with an average size which pratically do not change their size (peak in the distribution).

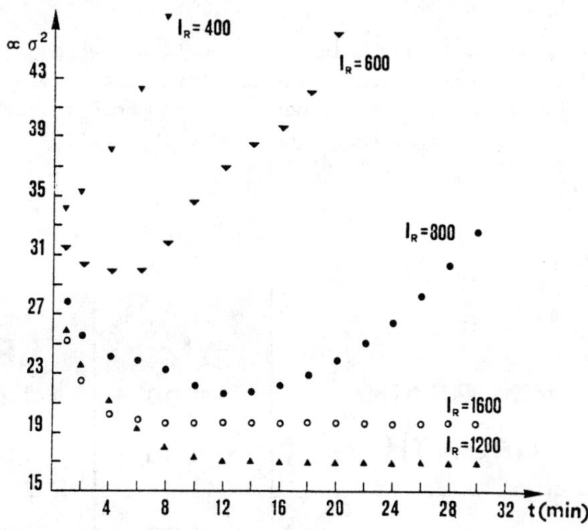

Figure 6 - Variance of the grain size distribution as a function of the time for different inhibition values

In Fig. 2 it is also possible to see that for $I_R \geqslant 400\ cm^{-1}$ the average grain grows, after a certain time, with a nearly linear law. Moreover for I_R in the range 800-1000 cm^{-1} the mean grain radius appears to be constant although the distribution is developing a long tail (see $R_{MAX}/\bar{R}$ ratio) which reflects a very slow growth in the mean radius. This corresponds to an incubation time which procedes sudden secondary recrystallization as soon as the few growing grains are able to consume all the other grains around the high peak in the size distribution. This is shown more clearly in Figs. 5 and 6, where $R_{MAX}/\bar{R}$ and σ^2 are plotted versus time. It is worth noticing that in the time interval of 15' and for the inhibition range 800-1000 cm^{-1} the size distribution develops the maximum tail and this corresponds to an increase of the variance (Fig. 6). Instead when the inhibition intensity is higher the size distribution goes to an equilibrium, changing its shape and reducing the dispersion around the average radius. For a very high value of I_R the change can be very feeble and this leads at the same time to a small change of the initial variance (compare in Fig. 6 I_R = 1200 with I_R = 1600).

In general, secondary recrystallization occurs in the inhibition range for which $2/3 < I_R R_c < 1$ is valid. In this range secondary recrystallization develops gradually from a more continuous feature (for the $I_R R_c = 2/3$) towards a more discontinuous one for $I_R R_c = 1$.

It is interesting to point out that these conditions were already established by Hillert (1) (see Fig. 7) starting from some topological considerations in a bidimensional defect model. Hence these conditions seem to be intrinsic to the statistical model where the critical radius concept is used.

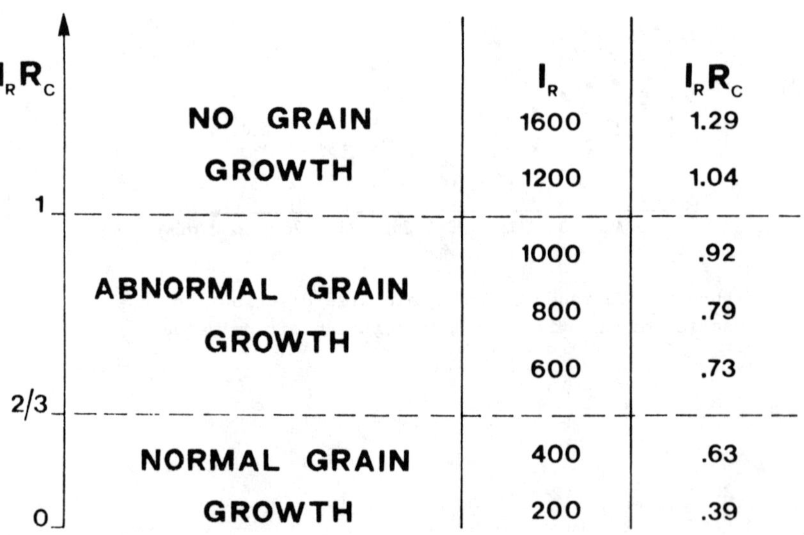

I_R	$I_R R_C$
1600	1.29
1200	1.04
1000	.92
800	.79
600	.73
400	.63
200	.39

Figure 7 - $I_R R_c$ products after 30 minutes of simulation time in comparison with the Hillert scheme (1)

Moreover if a special normalized plot for the size distribution is used (9) when the condition $I_R R_c = 1$, the grain size distribution seems to be independent of the initial conditions. That is to say, when $I_R R_c = 1$ the size distribution has lost all "memory" of the shape of the initial distribution and assumes a form which appears to completely describe the topological circumstances able to realize a sudden abnormal growth (a very high probability of contact among the few big grains and a very large number of small grains).

In the framework of the statistical model it is also possible to calculate the analytical expression of the "critical" grain size distribution and to emphasize all the conditions that must be fulfilled to enable secondary recrystallization to occur. Because of lack of space, results will be published later on and only some general conclusions will be established here.

Another point is that the computer simulations shown above refer to a constant value of inhibition but it is possible to get almost the same results and characteristic features of the abnormal growth process simulating a drop of the inhibition from a high value (which leads to stagnation condition) to a critical value that permits the developing of a long tail in the size distribution.

Discussion on the conditions necessary for secondary recrystallization process

As already shown above, to get secondary recrystallization in a homogenous grain sized matrix, the presence of second phase particles or in general of Zener drag is necessary (and this is the only way if there is not any texture effect (5)). We will now discuss in more detail the case where the initial grain size distribution, in the presence of a certain amount of second phase particles, has achieved the condition of grain growth stagnation. On the other hand we want to show that as a consequence of a drop in the inhibition intensity (coarsening or dissolution of the particles), the grain size distribution change in a continuous manner adjusting itself to a new stagnation condition till a well defined critical limit condition, where some grains can grow suddenly, is achieved (abnormal grain growth in a homogenous matrix). The situation described by equation (6) and (7) for the grain size distribution is schematically shown in Fig. 8 where the stagnation condition is achieved when the range of validity of equation (7) is extended from the grain of minimum radius to the grain of maximum radius present in the system. The grain size distribution continuously "follows" the decrease in inhibition by continuously changing the characteristic parameters ($\bar{R}$, σ^2, R_{MIN}, R_{MAX}). In fact, by decreasing the inhibition intensity by a certain amount, both a well defined number of larger grains can grow and another fraction of smaller ones can shrink till disappearance from the system (see equation (6)). Contemporaneously this process increases the critical radius (8) of the size distribution constraining again all the grains to satisfy equation (7). The critical condition for which an equilibrium grain size distribution cannot be achieved and some larger grains can grow at the expense of the matrix grains is the following: the total driving force acting on a hypothetical large R_{MAX} growing grain must be, at a certain time, equal to the driving force acting on the critical radius in the opposite direction.

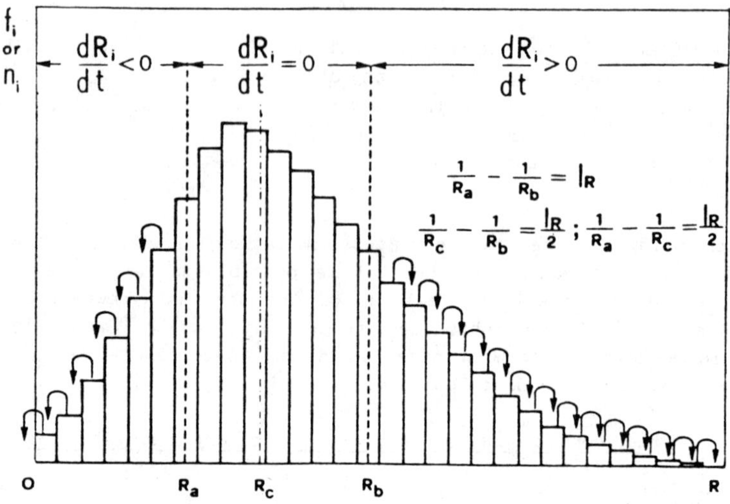

Figure 8 - Schematic representation of the flux of grains in the histogram in a certain interval Δt. In every instant t there are two grain sizes R_a and R_b satisfying relationships, shown in the figure, with the critical radius R_c and the intensity of inhibition I_R. When grain growth stagnation is established, R_a and R_b become equal to R_{MIN} and R_{MAX}, respectively. In this latter case, for all the grains the conditions dR/dt = 0 must be assumed.

To explain the last statement we have to consider some aspects of the equilibrium condition in terms of the driving forces. For complete grain growth stagnation condition equation (10b) can be rewritten for grains with $R = R_{MAX}$ and R_{MIN}:

$$\frac{1}{R_c} - \frac{1}{R_{MAX}} - \frac{I_R}{2} = \frac{1}{R_c} - \frac{1}{R_{MIN}} + \frac{I_R}{2} = 0 \qquad (11)$$

In Fig. 9 a radius R^+, greater than R_{MIN} and smaller than R_c is shown for which the following relationship is valid:

$$\frac{1}{R_c} - \frac{1}{R_{MAX}} - \frac{I_R}{2} < \frac{1}{R_c} - \frac{1}{R^+} + \frac{I_R}{2}$$

where the term on the right side is proportional to the energy that is necessary to overcome the restraining force which prevents the R^+ grain from shrinking.

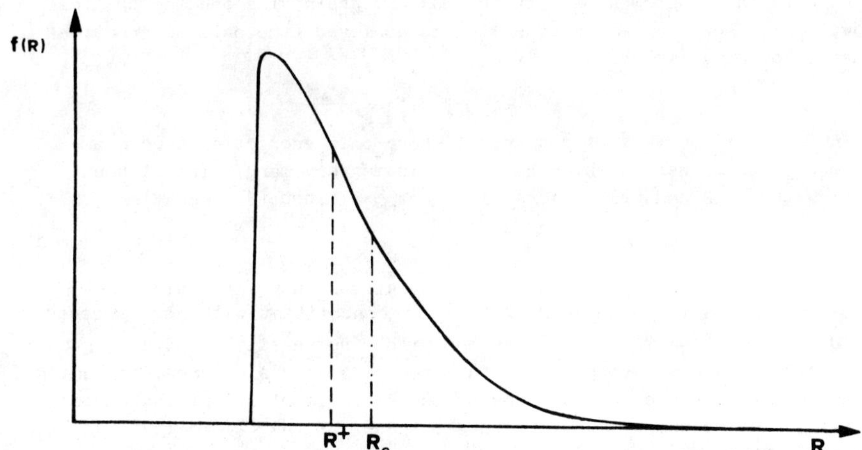

Figure 9 - Schematic drawing indicating the limiting conditions for abnormal grain growth. To achieve the critical stagnation condition the size distribution changes its shape to diminish the R_c/R_{MIN} and $R_c/\bar{R}$ ratios.

If now the limiting condition, $R^+ \longrightarrow R_c$, is considered it follows that:

$$\frac{1}{R_c} - \frac{1}{R_{MAX}} - \frac{I_R}{2} < \frac{I_R}{2} \qquad (12)$$

where on the right side $R^+ = R_c$ has been inserted into the previous equation.

The previous "critical" proposition states that when R_{MAX} has grown ($R_{MAX} \longrightarrow \infty$) equation (12) becomes an equality. That is, to start secondary grain growth the maximum grain radius, once large enough, must be subjected to a driving force able to overcome the excess of restraining force which "protects" the critical radius from shrinking:

$$\frac{1}{R_c} - \frac{I_R}{2} = \frac{I_R}{2} \qquad (13)$$

or $$I_R R_c = 1$$

In this way, starting from the stagnated condition a decreased inhibition results in the growth of both the maximum grain size and the critical radius, till a new stagnation condition is achieved (the only alternative is just abnormal growth).

In this limited type of reasoning, where only energetic aspects are considered, the manner in which R_{MAX} grows to very large size without affecting the mean and critical radii is not explained (see equation (13)).

As discussed above, the computer simulations show the rapid development of a long tail in the size distribution is concomitant with the presence of a high peak at small grain sizes and with a not easily decectable growth of the critical and mean radius. In fact the biggest grains grow, consuming the smallest grains due to the shape of the size distribution. The number of small grains consumed is rather large, therefore the volume involved is enough to allow the very few R_{MAX} grains to achieve a considerable size. On the other hand the mean radius which is the result of an average process $\sum_i f_i R_i$) is not particularly affected by the lost of the smallest grains.

Equation (13) says also that in the stagnation condition, just before the starting of the sudden secondary recrystallization, $1/R_{MAX} = I_R/2$ (equation (11)) must be valid. That is to say, the grain of maximum radius must have the growth pressure (1/R) just equal to the recstraining force. In Fig. 10 the nergetic terms in the critical stagnation conditions are shown.

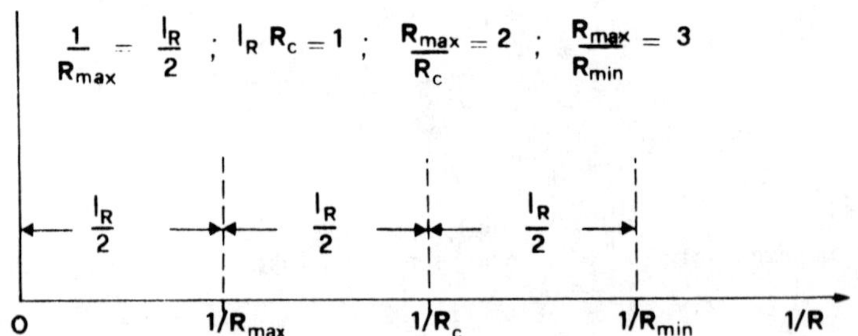

Figure 10 - Schematic representation to explain the condition on driving force to get secondary recrystallization. The limiting conditions on R_{MAX}/R_C and R_{MAX}/R_{MIN} come out putting together equations (10) and (13)

It is opportune to notice how condition (12) tells us nothing about the kinetics of the abnormal growth process but from the structure of equation (6) it is possible to recognize that starting from the stagna tion condition:

$$\frac{dR_{MAX}}{dt} = M \left(\frac{1}{R_c} - \frac{1}{R_{MAX}} - \frac{I_R}{2}\right) = 0$$

the growth of the R_{MAX} grains will be initially slow (incubation time) and then it will be followed by a sudden growth of these grains which at the most can procede with a linear law. Moreover, in this case they grow with the maximum velocity; in fact when R_{MAX} is large enough and is growing in a homogenous matrix of small grains (R_c sized) the following equation is valid:

$$\frac{dR_{MAX}}{dt} = M \left(\frac{1}{R_c} - \frac{I_R}{2}\right) \simeq \text{const}$$

and it is valid also for all the time that R_c changes very little during the growth of R_{MAX} grains. However it is established that grain growth is stopped when $I_R R_c = 2$

Another series of long time computer simulations are now under way to completely assess the characteristics of secondary recrystallization. Results will be published when complete; however it is possible to anticipate that, in the critical range of $I_R R_c \lesssim 1$, inside the characteristic long tail a second peak will appear which should precede the complete development of a bimodal size distribution till the complete disappearance of the major peak at smaller size. This is quite similar to the feature of the secondary recrystallization induced by texture as already established (5).

Conclusions

In the framework of the statistical model it appears possible to completely describe the main features of the normal and abnormal grain growth in a random matrix in presence of an inhibitor.

The characteristic of secondary recrystallization process, which appears evident in the range $2/3 < I_R R_c < 1$ are: a developing of a long tail in the grain size distribution; the presence of a high peak at smaller grain size; a linear law for the mean radius growth, at least for a wide range of time.

When grain growth stagnation is achieved ($I_R R_c > 1$) and the inhibition intensity begins to drop (due for istance to Ostwald ripening or dissolution of the particles) the grain size distribution modifies itself to achieve the critical condition ($I_R R_c = 1$) in which the shape of the size distribution is independent of the initial conditions.

References

1. M. Hillert, "On the Theory of Normal and Abnormal Grain Growth", Acta metallurgica, 13 (1965) 227-238.

2. P. Feltham, "Grain Growth in Metals" Acta Metallurgica, 5 (1957) 97-105.

3. N.P. Louat, "On the Theory of Normal Grain Growth", Acta Metallurgica, 22 (1974) 721-724.

4. O. Hunderi, N. Ryum and H. Westengen, "Computer Simulation Grain Growth", Acta Metallurgica", 27 (1979) 161-165.

5. G. Abbruzzese, K. Lücke, "A Theory of texture Controlled Grain Growth", Acta Metallurgica, to be published

6. T. Gladman, "On the Theory of the Effect of Precipitate Particles on Grain Growth in Metals", Proc. of the Royal Society London, 294A (1966) 298-309.

7. T. Gladman, F.B. Pickering, "Grain Coarsening of Austenitie", Journal of the Iron and Steel Institue, 205 (1967) 653-664.

8. P. Hellman, M. Hillert, "On the Effect of Second Phase Particles on Grain Growth", Scand. J. Metall, 4 (1975) 211-219.

9. G. Abbruzzese, "Computer Simulated Grain Growth Stagnation", Acta Metallurgica, 33 (1985) 1329-1337.

THE MICROSTRUCTURAL DYNAMICS OF PRIMARY AND SECONDARY RECRYSTALLIZATION

M. P. Anderson*, G. S. Grest* and D. J. Srolovitz#

*Corporate Research Science Laboratory
Exxon Research and Engineering Company
Annandale, NJ 08801

#Theoretical Division
Los Alamos National Laboratory
Los Alamos, New Mexico 87545

Monte Carlo computer simulation techniques have been applied to the problem of recrystallization in a two dimensional matrix. Both primary and secondary recrystallization are investigated. Primary recrystallization is modelled under conditions where the degree of stored energy is varied and nucleation occurs either continuously as a function of time or as site saturated. The degree of stored energy is adjusted to range from heterogeneous nucleation at grain edges and corners to homogeneous nucleation throughout the microstructure. Secondary recrystallization is modelled: (a) where the driving force is provided solely by curvature and (b) where the driving force is provided by the difference in the solid-vapor surface energy between grains of different crystallographic orientation. The role of particles in initiating secondary recrystallization is also addressed. The microstructure and kinetics for the different recrystallization processes are discussed and compared.

Introduction

The complete prediction of microstructural development in polycrystalline solids as a function of time and temperature is a major objective in materials science, but has not yet been possible primarily due to the complexity of the grain interactions. The evolution of the polycrystalline structure depends upon the precise specification of the coordinates of the grain boundary network, the crystallographic orientations of the grains, and the postulated microscopic mechanisms by which elements of the boundaries are assumed to move. Therefore, a general analytical solution to this multivariate problem has not yet been developed. Recently, we have been able to successfully incorporate these aspects of the grain interactions, and have developed a computer model which predicts the main features of the microstructure from first principles (1,2). The polycrystal is mapped onto a discrete lattice by dividing the material into small area (2d) or volume (3d) elements, and placing the centers of these elements on lattice points. This discrete model preserves the topological features of real systems, and can be studied by computer simulation using Monte Carlo techniques. The application of the model to normal grain growth with isothermal annealing is presented elsewhere in this volume (3). In this paper we will review extension of the model to treat primary and secondary recrystallization.

Primary recrystallization is the process in which a deformed metal is transformed into a strain-free structure by the nucleation and growth of strain-free grains (4). The dominant driving force for this phenomenon is assumed to be excess energy distributed in the interior of the deformed grains, in the form of point defects and dislocations. Experimentally it is often observed that the recrystallized volume fraction F exhibits a sigmoidal dependence on time (5). This type of dependence can often be fit by a simple mathematical relationship derived by Johnson and Mehl (6), and Avrami (7) (JMA):

$$F = 1 - \exp(-At^p) \qquad (1)$$

where A and p are constants. An important difference which exists between recrystallization and normal grain growth is that the motion of the recrystallized boundaries is dominated by the volume (3d) or area (2d) distributed energies. As a first attempt to directly correlate microstructural development with this type of driving force, the temporal evolution of microstructure is examined as a function of the nucleation rate and stored energy density. Primary recrystallization is modelled under conditions where the degree of stored energy is varied and nucleation occurs either heterogeneously or homogeneously. The nucleation rate is chosen as either constant or site saturated. Details of the nucleation mechanism are not, however, investigated in the present study.

Grain growth is the term used to describe the increase in grain size which occurs upon annealing after primary recrystallization is complete. At least two different types of grain growth phenomena have been distinguished. Normal grain growth is said to occur when the microstructure exhibits a uniform increase in size (8). This process has been treated in a companion paper (3). The second type of growth which can take place is secondary recrystallization. In the present context, secondary recrystallization refers to the rapid increase in size of a few grains in the recrystallized microstructure such that the maximum grain size increases at a rate much faster than the arithmetic mean (9). This type of growth is similar to primary recrystallization in that the volume fraction of secondary grains often exhibits sigmoidal time dependence, and can be fit by the JMA relationship (Eq.1). At least two different origins for secondary recrystallization have been suggested. In the first it is assumed that the driving force for growth is derived solely from the decrease in total grain boundary energy (curvature) (10-12). Secondary recrystallization then occurs as a consequence of growth instabilities predicted under a restricted set of conditions by mean field treatments of grain growth. The second explanation postulates that, in addition to curvature, a volume (3d) or area (2d) distributed excess energy must be present (13-15). Secondary recrystallization then initiates as a consequence of local differences in the volume distributed energy. For 2d materials (sheets and thin films), experimental evidence indicates that one source of this additional driving force is the orientation dependence of the solid-vapor surface energies for the recrystallized grains. In an attempt to check the validity of these alternatives, the growth of abnormally large grains in a two-dimensional matrix is modelled under two conditions: (a) where the driving force is provided solely by curvature and (b) where the driving force is provided by the difference in the solid-vapor surface energy between grains of different crystallographic orientation.

Procedure

The procedures employed in the simulations are described in Ref. 16 for primary recrystallization, and Ref. 17 for secondary recrystallization. In short, a continuum microstructure is mapped onto a two dimensional triangular lattice containing 40,000 lattice sites. Each lattice site is assigned a number, S_i, which corresponds to the orientation of the grain in which it is embedded. The number of distinct grain orientations is Q. Lattice sites which are adjacent to neighboring sites having different grain orientations are regarded as being part of the grain boundary, whilst a site surrounded by sites with the same grain orientation is in the grain interior.

The site energy, E_i, is defined to be the sum of two contributions, E_i^{I} and E_i^{II}. The grain boundary energy is specified by E_i^{I}

$$E_i^{I} = - J \sum_{nn} (\delta_{S_i S_j} - 1) \tag{2}$$

where $\delta_{i,j}$ is the Kronecker delta, the sum is taken over the nearest neighbor (nn) sites and J is a positive constant that sets the scale of the grain boundary energy. E_i^{II} specifies the energy stored within the grain interior due to deformation or (in 2d) to the solid-vapor surface energy.

To treat primary recrystallization, E_i^{II} is defined to be a positive energy such that

$$E_i^{II} = H\,\theta(Q_u - S_i) \tag{3}$$

where $\theta(x) = 1$ for $x \geqslant 0$ and $= 0$ for $x < 0$. H is a positive constant which sets the magnitude of the stored energy and Q_u is the number of distinct orientations of unrecrystallized grains. Strain-free grains are nucleated in the microstructure under conditions described in the results section below. These recrystallized nuclei are given orientations with $S_i > Q_u$, such that the S_i for a given recrystallized nucleus is shared by no other grains or nuclei.

To treat secondary recrystallization, E_i^{II} is defined to be a negative energy such that

$$E_i^{II} = -H \sum_k \delta_{S_i\, Q_k} \tag{4}$$

H is a positive constant which sets the magnitude of the solid-vapor surface energy. In simulations of secondary recrystallization where curvature is used as the only driving force, H=0. For those simulations where surface energy is employed as an additional driving force, H>0. If, in a given annealing atmosphere, certain grains have a lower total energy than surrounding grains due to the dependence of surface energy on the crystallographic orientation of the surface plane, it is speculated that these grains will grow faster - resulting in secondary recrystallization (13-15). H scales the amount by which the surface energy of a grain having orientation S_i is lowered. The summation in Eq. 4 is over the set of orientations Q_k, which is a list of favored grain orientations chosen at random at the start of the simulation from the Q total orientations of the recrystallized microstructure. The delta function in Eq. 4 is one if S_i in one of the favored orientations, and zero if it is not.

The kinetics of the boundary motion are simulated employing a Monte Carlo procedure in which a lattice site is selected at random and its orientation is randomly changed to one of the other grain orientations. The change in energy, ΔE, associated with the change in orientation is evaluated. If the change in energy is less than or equal to zero, the re-orientation is accepted. However, if the change in energy is greater than zero, the re-orientation is accepted with the probability $\exp(-\Delta E/kT)$. Time, in these simulations, is related to the number of re-orientation attempts. N re-orientation attempts is arbitrarily used as the unit of time and is referred to as 1 Monte Carlo Step (MCS), where N is the number of lattice sites (40,000). The conversion from MCS to real time has an implicit activation energy factor, $\exp(-W/kT)$, which corresponds to the atomic jump frequency. Since the quoted times are normalized by the jump frequency, the only effect of choosing T=0 (as done in the present simulations) is to restrict the accepted re-orientation attempts to those which lower the energy of the system. Re-orientation of a site at a grain boundary corresponds to boundary migration. The boundary velocity determined in this manner yields kinetics that are formally equivalent to the rate theory model.

Results and Discussion

A. Primary Recrystallization

In classical homogeneous nucleation theory the minimum radius of a spherical nucleus which will grow is $2\gamma/\delta G$, where γ is the surface energy per unit area and δG is the difference between the energy stored in the nucleus and that stored in the surrounding matrix (normalized per unit volume). In the present simulations, the minimum nucleus radius which will grow but cannot shrink without regard to location in the microstructure is proportional to J/H, where the proportionality constant depends upon the nucleus shape. While the critical nucleus size in a continuum system is a continuous function of J/H, on a discrete lattice at $T \cong 0$ the critical size changes discontinuously. On a triangular lattice the critical nucleus size is one site for H/J>4, three sites for 4>H/J>2, and essentially infinite for 2>H/J>0. For H/J<2, the nucleation can only be heterogeneous.

For metallic systems undergoing recrystallization, the ratio H/J is in the range 0.01 to 0.1 (18). In the present simulations H/J was selected to range from 0.1 to 5.0. Although these values are in excess of those observed in metals, they are required in the discrete lattice model to vary between heterogeneous to homogeneous nucleation and growth behavior.

1. Circular Grains

The behavior of individual, growing, recrystallized grains in a uniform background was investigated first. The temporal evolution of the recrystallized grains for two values of stored energy, H/J=3 and H/J=5, are shown in Fig. 1. Three significant features are found. First, the grains

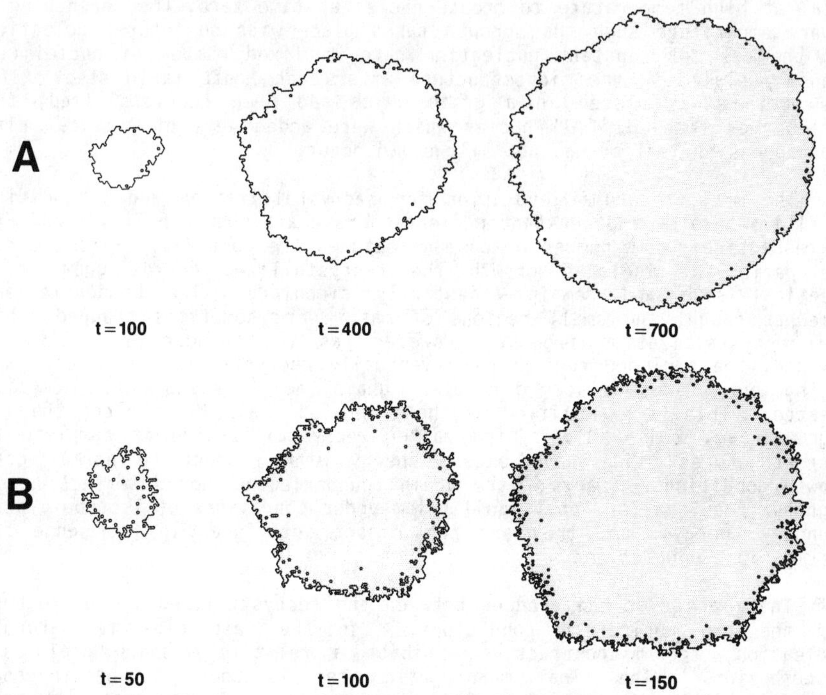

Fig. 1 The growth of a circular grain in a uniform background (a) H/J=3 and (b) H/J=5.

evolve such that the grain shape remains nearly circular. This indicates that the grain boundary mobility is isotropic and insensitive to the symmetry of the underlying lattice. Second, the rate of growth increases asymptotically with time squared. This is consistent with exact analytical solutions for this simple geometry, and verifies that the boundary migration kinetics are properly incorporated. Finally, the grain boundaries are not perfectly smooth, but exhibit a small degree of roughness. The magnitude of the roughness increases with stored energy. This is regarded as a finite size effect in the model, which is due to the discreteness and low dimensionality of the simulation. The scale of roughness does not increase with the size of the grain, and hence the normalized roughness decays with increasing domain size.

2. Homogeneous Nucleation

The recrystallization of a polycrystalline matrix under conditions of homogeneous nucleation was investigated next. By homogeneous nucleation is meant that the degree of stored energy is high enough that an embryo of critical size will grow in a uniform background. For these simulations, the microstructure was initialized by a 1000 MCS run of a Q_u=48 normal grain growth model, following which the clock was reset to zero. The ratio of stored energy to grain boundary energy, H/J, was chosen to be either 3 or 5. Two types of nucleation were considered: site saturated and constant nucleation rate. For site saturated nucleation, a fixed number of nuclei were randomly placed in the microstructure and no additional new nuclei were created. Physically, this is equivalent to annealing the deformed metal at high temperature to create nuclei at time zero, then quenching to lower temperature such that growth takes place with no future nucleation. For the case of constant nucleation rate, a fixed number of nuclei were randomly placed in the microstructure after each Monte Carlo step. If a new nucleus was placed in a grain which had been recrystallized, that nucleus was removed. All nuclei which were added were of critical size. Spontaneous nucleation was not allowed to occur.

The microstructural evolution for recrystallization under conditions of site saturated and constant nucleation rate are shown in Fig. 2 and Fig. 3, respectively. A number of common features are observed. First, during the period of isolated growth the recrystallized grains grow at an identical rate and remain essentially circular. The boundaries are somewhat rough, and small regions of matrix are sometimes trapped behind the recrystallization fronts. However, as the boundaries continue to advance, the occluded regions are eventually recrystallized. Second, when impingement takes place irregular grain morphologies are sometimes created. This is exemplified by the elongated grains seen along the left edge of Fig. 2 (t=820). Third, after recrystallization is complete the microstructures still have excess energy with respect to normal grain growth conditions. Many of the grain boundaries do not intersect at 120 degrees (required for local equilibrium under conditions of isotropic grain boundary energy), and they are non-compact interfaces in the sense that they exhibit undulations.

There are also differences between the recrystallized microstructures for the two nucleation conditions. In the case of site saturated nucleation, the microstructure exhibits a relatively sharp grain size distribution. The final mean grain area is found to be inversely proportional to the initial nuclei concentration. In contrast, the grain size distribution obtained from a constant nucleation rate is broader due to continued nucleation of recrystallized grains as a function of time. The microstructure is topologically different in that it contains two sided grains, as shown on the right side of Fig. 3 (t=800). These grains occur when nucleation takes place between two larger pre-existing grains (a condition which is never met for site saturated nucleation). The final mean grain area is found to be inversely proportional to the nucleation rate to the 2/3 power. This is consistent with theoretical predictions by Gilbert (19) for a randomly nucleated microstructure.

The recrystallized volume fraction F was followed as a function of time, and is shown in Fig. 4 for site saturated nucleation in which H/J=3. Under all nucleation conditions, F is found to be a sigmoidal function of time, in agreement with experimental observation. The simulations show no incubation period for recrystallization. However, in the case that the experimentally detectable recrystallized grain size is finite, an apparent incubation period will exist.

The Avrami exponent p was evaluated for the different nucleation conditions from

$$p = \frac{\partial \log (\log (1/(1-F))}{\partial \log t} \tag{5}$$

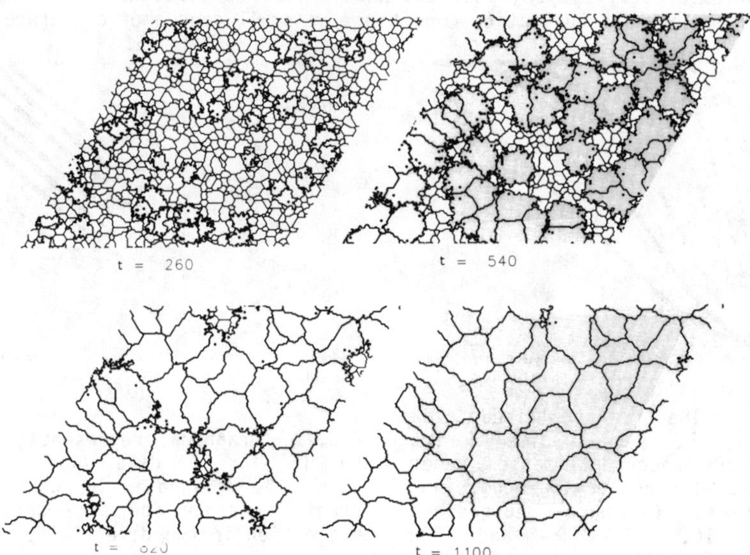

Fig. 2 The temporal evolution of the microstructure during homogeneous recrystallization under site saturated nucleation conditions (50 nuclei/40,000 sites) with H/J=3. The shaded grains have been recrystallized.

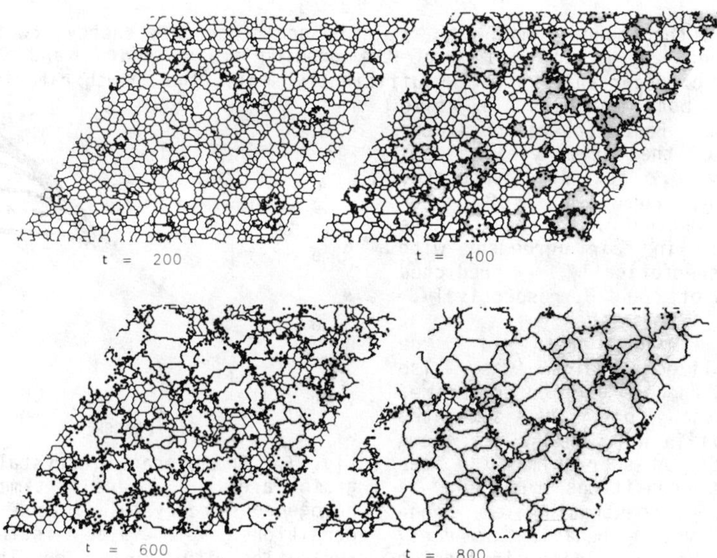

Fig. 3 The temporal evolution of the microstructure during homogeneous recrystallization under constant nucleation rate conditions (0.2 nuclei/MCS) with H/J=3. The shaded grains have been recrystallized.

Such a determination is shown in Fig. 5 for the data of Fig. 4. If the JMA equation is exactly obeyed, the data should fall on straight lines. It is found that, for both nucleation conditions, a small degree of curvature is

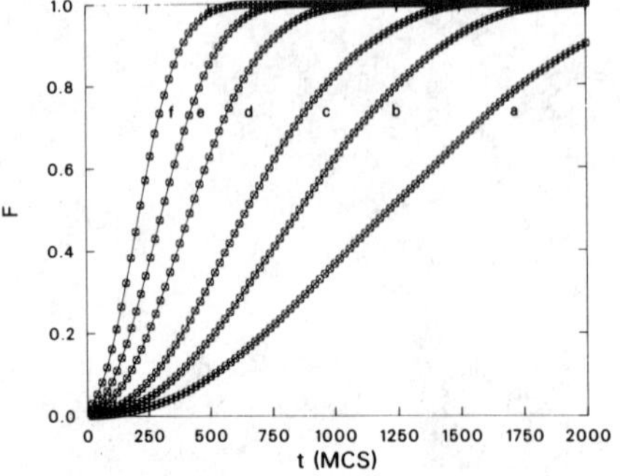

Fig. 4 The recrystallized area fraction, F, versus time during homogeneous recrystallization under site saturated nucleation conditions with H/J=3. Curves a-f correspond to 5, 10, 50, 100 and 200 nuclei/40,000 sites, respectively.

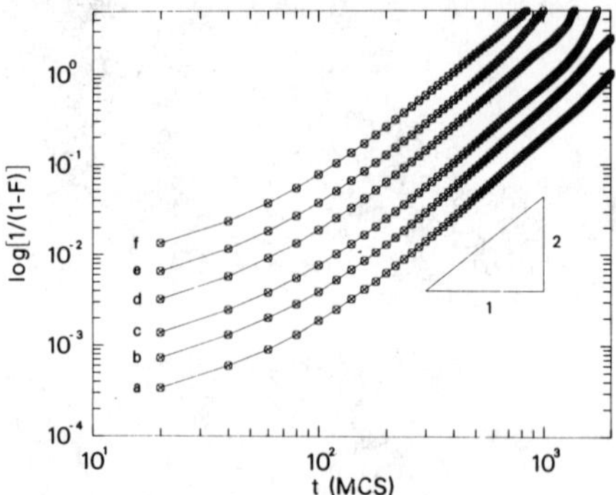

Fig. 5 Graphical representation of Eq. (5) for the data shown in Fig. 4. The triangle indicates the predicted slope for site saturated nucleation in two dimensions.

present, particularly at very short times. It is possible that the short time deviation is due to the departure from the linear relationship between recrystallized grain radius and time at small grain sizes. Theoretically, the growth rate for an individual recrystallized grain is given by

$$\frac{dR}{dt} = M\left(-C_1\gamma/R + C_2\,\delta G\right) \qquad (6)$$

where M is the boundary mobility, γ is the grain boundary energy, δG is the stored energy R is the radius of the growing grain, and C_1 and C_2 are constants of order unity. For R sufficiently small, the growth rate is not constant but exhibits a size dependence. Best fits to the linear portions of the curves yield slopes of 2.3 ± 0.2 for site saturated nucleation, and 3.0 ± 0.2 for constant nucleation rate. These values are in fair agreement with the theoretically predicted exponents of 2 and 3, respectively.

The mean area of the recrystallized grains was also followed as a function of time. These data for site saturated nucleation in which H/J=3 are shown in Fig. 6. At short times for both nucleation conditions curvature is observed, demonstrating the non-linear R vs. t behavior at early times. However, with increasing time, linear behavior ensues, such that A varies with t^2 as expected. At later times the mean area saturates due to total grain

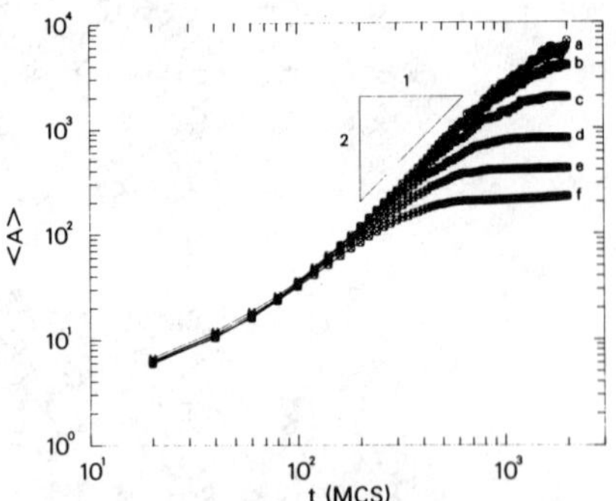

Fig. 6 The average recrystallized grain area, A, versus time for homogeneous recrystallization under conditions of site saturated nucleation with H/J=3. The slope of 2 indicated by the inset triangle represents the growth kinetics of a recrystallizing grain in a uniform background.

impingement. The slope of A at saturation is much less than one, indicating that normal grain growth has not yet begun. Presumably, local equilibration of vertices and boundary smoothing must take place before normal grain growth can begin.

3. Heterogeneous Nucleation

The recrystallization of a polycrystalline matrix under conditions of heterogeneous nucleations was also investigated. By hetergeneous nucleation is meant that the degree of stored energy is too small to support growth in a uniform background for an embryo of given size. Successful nucleation (growth of the embryo) can only take place at pre-existing heterogeneities in the microstructure. As before, the microstructure was initialized by a 1000 MCS run of a Q_u=48 normal grain growth model, following which the clock was reset to zero. The ratio of stored energy to grain boundary energy was chosen to be either 0.5, 1.0, 1.5, or 2.0. For these amounts of stored energy the critical embryo size required for homogeneous nucleation is essentially infinite. Two types of nucleation were considered: site saturated and constant nucleation rate. As before, for site saturated nucleation a fixed number of nuclei were randomly placed in the microstructure, and no additional nuclei were created. For the case of constant nucleation rate, a fixed number of nuclei were randomly placed in the mirostructure after each Monte Carlo step. If a new nucleus was placed in a grain which had been recrystallized, that nucleus was removed. The size of the nucleus employed was three lattice sites.

The microstructural evolution for recrystallization under conditions of site saturated and constant nucleation rate are shown in Fig. 7 and 8, respectively. A number of differences with respect to homogeneous nucleation are observed. First, only embryos which are present on grain boundaries or grain boundary intersections grow. At the lowest stored energy employed, successful nuclei are found exclusively on grain vertices. As the stored energy is increased, successful nucleation then begins to occur on the two-grain interfaces (grain boundaries). Embryos which are present in the grain interiors shrink and disappear. During growth, the boundaries of th recrystallized grains are smoother than for homogeneous nucleation, and occlusion of unrecrystallized matrix does not occur. Second, the growth rates of the recrystallized grains before impingement are not identical; some grains grow faster than others. This is most obvious in the case of site saturated nucleation (Fig. 7), where all nuclei start at time zero with the same initial size. Third, while impingement still leads to some irregular grain morphologies, the recrystallized grain shapes tend to be much more equiaxed than for homogeneous nucleation. Fourth, after recrystallization is complete the grain boundary configuation is much closer to the normal grain growth state than for homogeneous nucleation. The boundaries are much more compact in that they do not have the undulations found above, and many intersect at 120 degrees (satisfying local equilibrium).

There are also similarities with respect to homogeneous nucleation. The grain size distribution is sharper for site saturated nucleation in comparison to constant nucleation rate, as previously observed. The topological differences between site saturated and constant nucleation rate are still present (e.g. the occurrence two-sided grains).

The recrystallized volume fraction F was followed as a function of time, and is shown in Fig. 9 for constant nucleation rate in which H/J=0.5. Under both nucleation conditions, F is again found to be a sigmoidal function of time, except for site saturated nucleation at the

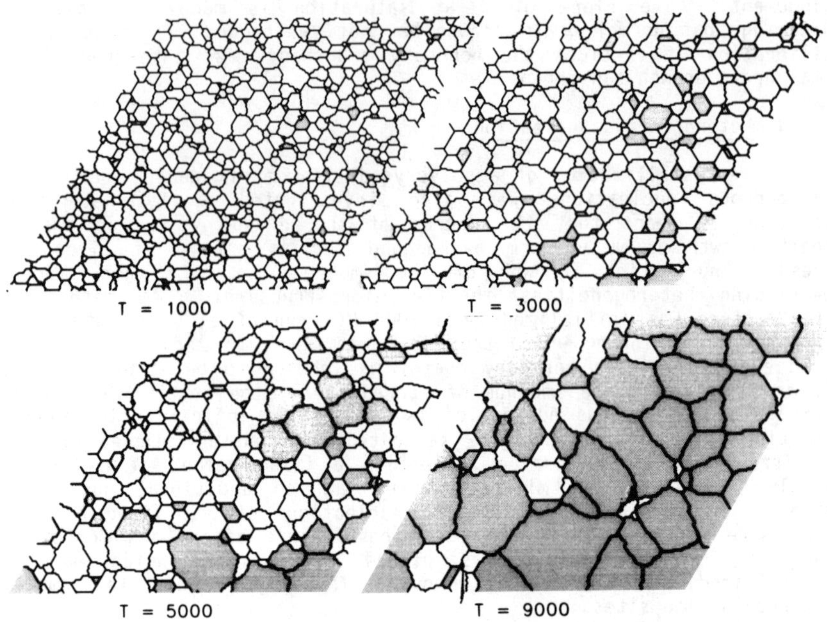

Fig. 7 The temporal evolution of microstructure during heterogeneous recrystallization under site saturated nucleation conditions (500 nuclei/40,000 sites) with H/J=0.5.

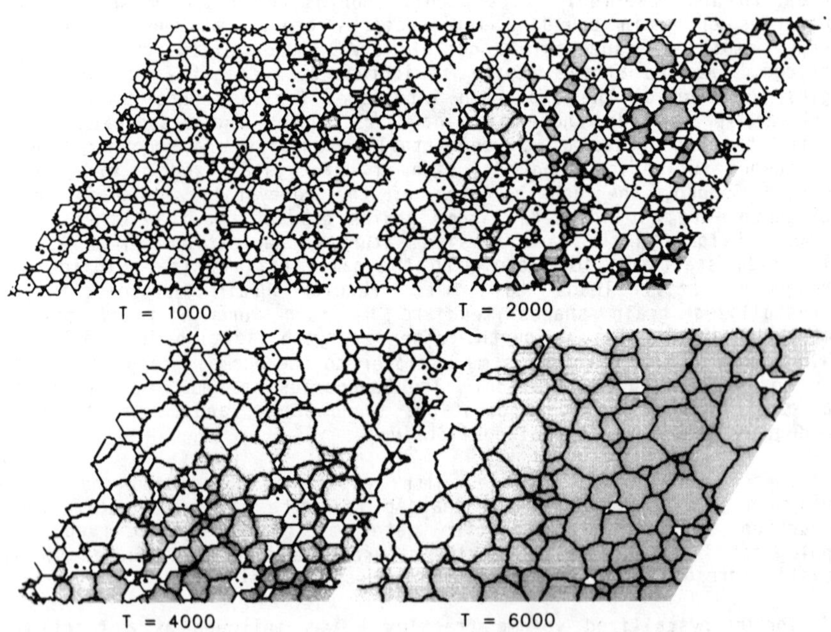

Fig. 8 The temporal evolution of microstructure during heterogeneous recrystallization under constant nucleation rate conditions (3 nuclei/MCS) with H/J=0.5.

lowest value of stored energy. In this case a small dip is observed at short times due to the elimination of unsuccessful embryos. At long times the sigmoidal behavior is recovered. As before, the simulations show no incubation period for recrystallization.

The recrystallized volume fraction was plotted in accordance with Eq. 5 to check the validity of the JMA analysis. These results are shown in Fig. 10 for the data of Fig. 9. Unlike for homogeneous nucleation,

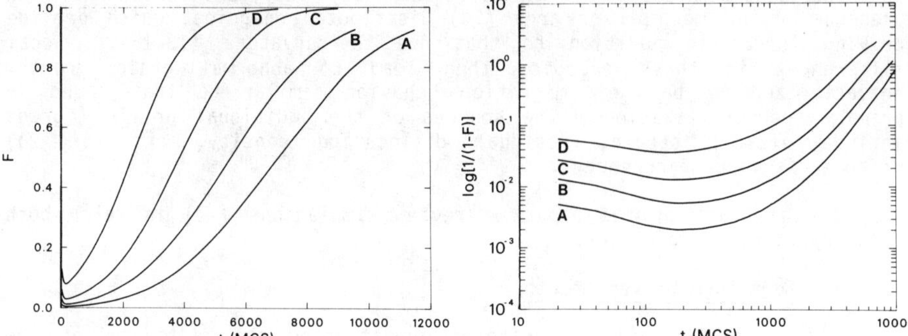

Fig. 9 The recrystallized area fraction, F, versus time during heterogeneous recrystallization under conditions of site saturated nucleation for H/J=0.5. Curves a-d correspond to 200, 500, 1000, and 2000 nuclei/40,000 sites, respectively.

Fig. 10 Graphical representation of Eq. (5) for the data of Fig. 9.

significant deviations from linearity are observed. No attempt was made to fit the data for an Avrami exponent. The substantial violation of linearity may be a consequence of the different growth rates exhibited by the recrystallized grains. In the JMA analysis, it is assumed that all grains grow at identical rates. The variable growth rate observed is thought to be a consequence of two factors. First, a non-linear dependence of growth rate on size beyond that predicted by Eq. 6. Second, a dependence of growth rate on details of the local environment at small size. Growth appears to occur by motion of the vertices of the recrystallized grain along pre-existing boundaries. Consequently, at small size the growth rate will be substantially affected by the number of vertices the embryo has.

B. Secondary Recrystallization

The occurrence of secondary recrystallization in a microstructure where primary recrystallization is complete has been attributed to a number of sources. These sources can be divided into two types by dominant driving force. The first possible source is traced to instabilities which are predicted to occur in grain growth where the driving force is provided solely by curvature (10-12). The growth instabilities are triggered in one of three ways:

(a) If an abnormally large grain (greater than 2R in 2d or 9/4 R in 3d) is introduced into the grain ensemble in a system free from grain growth restraints.
(b) If particle retardation of grain growth takes place but the Zener drag is small and the system has not completely pinned.

(c) If the microstructure is completely pinned by particles but a time dependent decrease in Zener drag (through a reduction in the number of particles by dissolution or Ostwald ripening) takes place.

The models used to make these predictions assume that grains are spherical and growing in an average environment.

The second possible origin for secondary recrystallization is the presence of volume (3d) or area (2d) distributed energies, which provide driving forces in addition to that due to curvature (13-15). Local differences in these energies then lead to abnormal grain growth, characterized by boundary migration behavior similar to that found in primary recrystallization. The sources of the additional driving forces include elastic strain, residual dislocation density, and (in 2d) differences in surface energy.

In this part of the paper we review simulations (17) in which both these hypotheses are tested.

1. Curvature Driven Growth

The first simulations performed investigated the stability of the grain size distribution function developed during normal grain growth. In this case H=0 in Eq. (4). A microstructure produced by a 1000 MCS run of a Q=48 normal grain growth model was modified by artificially creating a large circular grain in its center. Initial circular grain sizes of 5, 10, 15, and 20 times the mean grain radius was used. The resulting microstructural development is shown in Fig. 11 for the case where $R_0/\bar{R}$ is

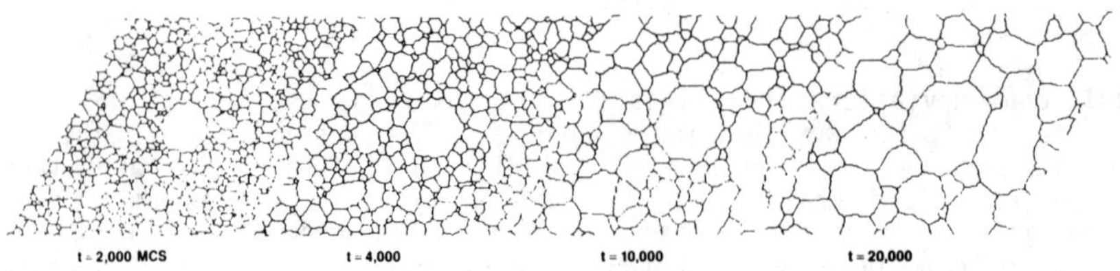

Fig. 11 The growth of a large grain in an otherwise normal grain growth microstructure.

initially 5. R_0 is the radius of the large circular grain, and $\bar{R}$ is the average grain radius excluding the large grain. The ratio of the area of the large grain, A_0, to the mean grain area excluding the large grain, $\bar{A}$, as a function of time is given in Fig. 12. These data show that the abnormal grain is a temporary feature of the microstructure. That is, the abnormal grain grows at a rate slower than the remainder of the microstructure until it is absorbed into the normal grain size distribution.

Since abnormally large grains do not of themselves lead to abnormal grain growth, simulations were performed in the presence of particle dispersions. In these simulations grain growth was allowed to occur in microstructures containing particles until stagnation took place. Following complete pinning, two different experiments were performed.

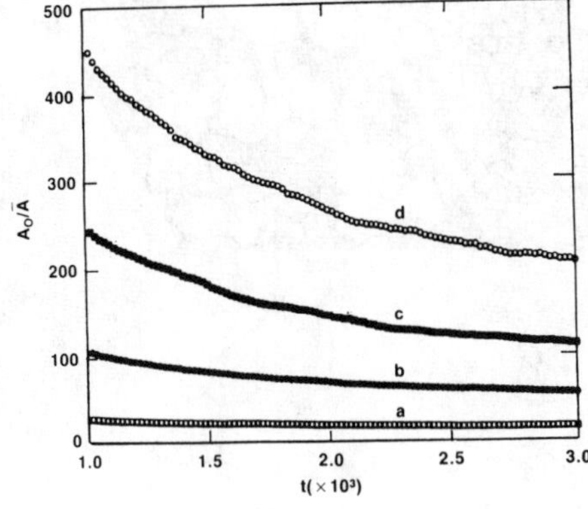

Fig. 12 The time dependence of the ratio of the areas of the large grain, A_0, to the mean grain area, $\bar{A}$. The data is presented for initial large grain radii (at 1,000 MCS) of (a) 5, (b) 10, (c) 15, and (d) 20 $\bar{R}$.

First, the concentration (area fraction) of particles was uniformly reduced from 2.5% to 1.5%, and the simulation was re-started. Second, all the particles were removed from one part of the microstructure to create a particle-free region, and the simulation was re-started. The time evolution of this microstructure is shown in Fig. 13.

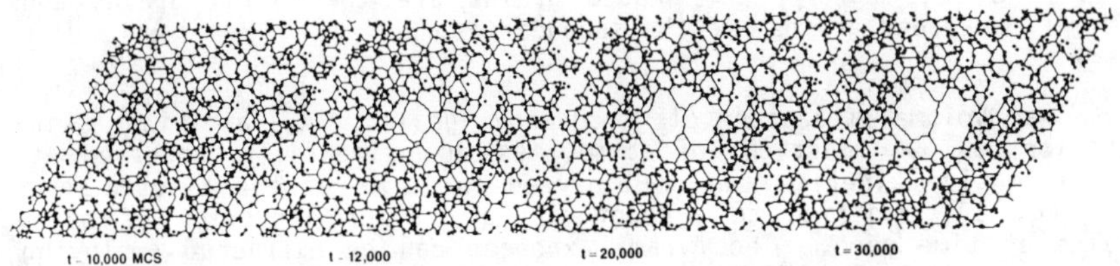

Fig. 13 Temporal evolution of a pinned microstructure following the removal of a large group of particles from one area.

For all these cases, while short growth transients were observed, pinning eventually took place and abnormal grain growth was not initiated.

2. Surface Energy Driven Growth

To test the second hypothesis concerning the origin of secondary recrystallization, simulations were carried out with H=0.1J in Eq. (4). In these studies, a normal grain growth microstructure produced by a 1000 MCS run of a Q=48 model was re-labeled so that each grain had a unique orientation. A fraction f of the grains was then selected as initial secondaries. f was chosen as either 2%, 5%, 10%, or 20%.

Figure 14 shows the microstructural evolution of a structure where initially 2% of the grains were secondaries. It is seen that the secondaries grow at a much more rapid rate than do the matrix grains, and that after a relatively short period of time the entire microstructure is composed of secondary grains. This behavior is similar to that of primary recrystallization discussed above. In contrast to the homogeneous nucleation recrystallization model, however, the secondary grain boundaries remain smooth and no occluded regions are formed. After secondary recrystallization is complete, the microstructure is much closer to the normal grain growth state in that the grain shapes are compact and the boundary intersections are close to 120 degrees.

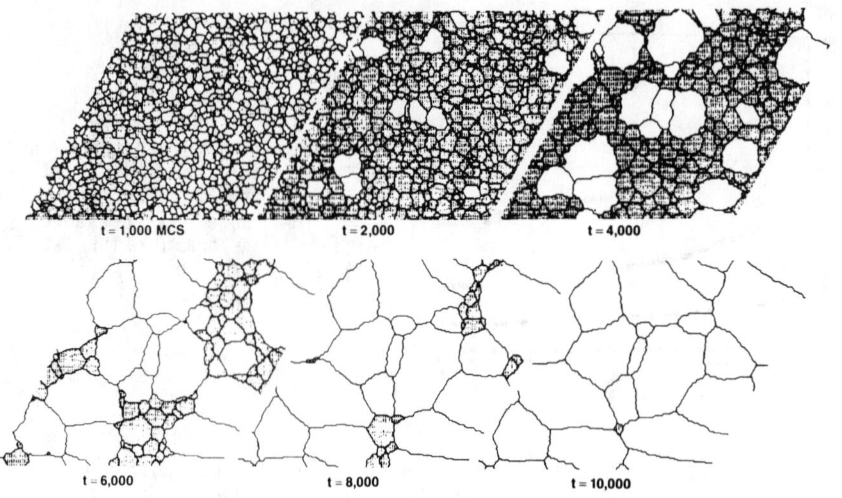

Fig. 14 Microstructure undergoing abnormal grain growth. In contrast to the microstructures shown for primary recrystallization (Fig. 2, 3, 7, and 8), the shaded grains are the matrix grains and the unshaded grains are the secondaries.

The volume fraction of secondary grains, X, exhibits sigmoidal behavior as a function of time, as experimentally observed. The transformation kinetics can be analyzed in terms of the kinetic model of JMA (Eq. 1), modified to account for a non-zero fraction of secondary grains at time zero. The Avrami exponent can be extracted employing the procedure of Eq. (5). This plot is shown in Fig. 15 for f = 2%, 5%, 10%, and 20%. Unlike the case of heterogeneous recrystallization model, small deviations from linearity occur. This is attributed in part to the larger initial size of the secondary grains at time zero as compared to the nuclei employed for heterogeneous recrystallization. As for high stored energy, it is expected that the growth rate becomes constant (size independent) when the recrystallized grain size is sufficiently large. By fitting the JMA equation to the data, an Avrami exponent $p = 1.8 \pm 0.3$ is obtained. This is close to the theoretical value of 2 expected for site saturated recrystallization, although for secondary recrystallization the assumption that all nuclei start growing from time zero with the same initial size is not fulfilled.

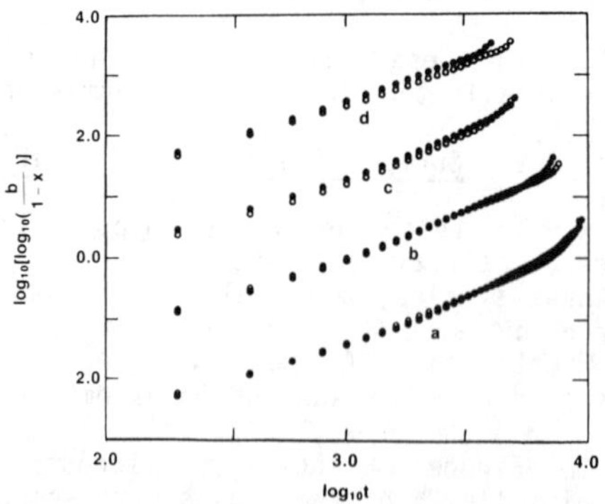

Fig. 15 Graphical representation of Eq. (5) for the fractional area of secondary grains, X, versus time. Curves a-d correspond to conditions in which 2%, 5%, 10% and 20% of the grains were initially secondaries. The quantity b is the initial area fraction of secondary grains.

Figure 16 presents the temporal evolution of the grain size distribution function during secondary recrystallization. At t=0, the grain size distribution function is just the normal grain growth distribution function. At intermediate times, the grain size distribution function consists of the large secondary grains and the much smaller matrix grains. This results in a broad grain size distribution, as experimentally observed. When the transformation is complete, the secondary grains compete against each other by normal grain growth and the size distribution assumes the same width as before the onset of abnormal grain growth.

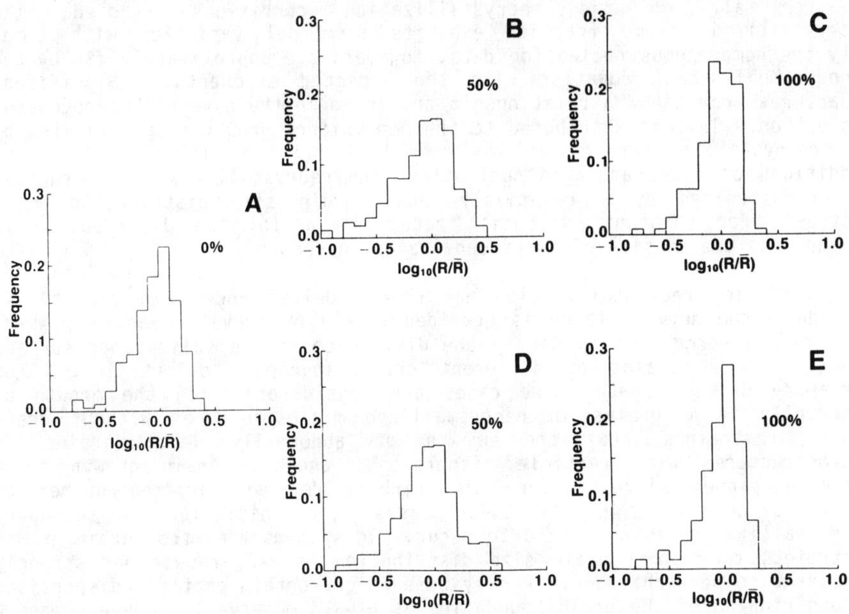

Fig. 16 The grain size distribution corresponding to (a) a normal grain growth microstructure at 1,000 MCS. Figures b and c correspond to the microstructure having initially (t=1,000 MCS) 2% secondary grains for X = 50% and 100%. Figures d and e are corresponding distributions for the microstructure having initially 10% secondary grains for X = 50% and 100%.

Summary

A simple and general simulation technique has been developed which easily lends itself to the analysis of complex microstructure. The technique has been applied to the problem of primary and secondary recrystallization.

Primary recrystallization has been modelled under conditions where the degree of stored energy is varied and nucleation occurs either homogeneously or heterogeneously. The nucleation rate is chosen as either constant or site saturated. As the degree of stored energy is increased, a number of changes take place in the characteristics of the recrystallized grains. First, the locations where successful nucleation occur increase in number to progressively include grain boundary vertices, two-grain interfaces (grain boundaries), and ultimately grain interiors. Second, the boundaries of the growing grains change from being relatively smooth to rough, with a greater tendency to trap unrecrystallized matrix behind the recrystallization front. Third, the growth rate for different recrystallized grains changes from being non-uniform to essentially identical. Fourth, the morphology of grains in the recrystallized microstructure changes from equiaxed to a greater irregularity of shape (elongated). The final grain boundary configuration departs increasingly from that of a normal grain growth state in that interfaces become non-compact (undulate) and boundary intersections increasingly deviate from 120 degrees.

For all of the recrystallization conditions studied, the recrystallized volume fraction exhibits sigmoidal variation with time. Only the homogeneous nucleation data, however, are approximately fit by the Johnson-Mehl-Avrami equation with the expected exponents. Significant departures from the JMA relationship are found in the case of heterogeneous nucleation. This is attributed to the non-uniform growth rate exhibited by the recrystallized grains at low stored energy and small size. Under conditions of site saturated nucleation, the recrystallized microstructure is characterized by a relatively sharp grain size distribution. In contrast, for constant nucleation rate, the grain size distribution is broader and the density of two-sided grains increases.

Secondary recrystallization has been modelled under two conditions: (a) where the driving force is provided solely by curvature and (b) where the driving force is provided by the difference in the solid-vapor surface energy between grains of different crystallographic orientation. For curvature driven growth three cases are considered: (a) the growth of abnormally large grains in recrystallized microstructures without grain growth restraints, (b) the growth of abnormally large grains in microstructures with particle dispersions, and (c) grain growth in a particle pinned microstructure in which a decrease in the number of particles occurs. In all these cases, the initiation of secondary recrystallization is not found to occur. In systems free from grain growth restraints the normal grain size distribution is very robust and strongly resistant to perturbations. For systems which contain particle dispersions strong pinning of the grain boundaries is always observed. However, when a preferred surface energy orientation is introduced, secondary recrystallization does take place. The microstructural evolution observed during secondary recrystallization is in good correspondence with experiment. The volume fraction of secondary grains exhibits sigmoidal behavior as a function of time, and is fit by the JMA equation.

References

1. M. P. Anderson, D. J. Srolovitz, G. S. Grest and P. S. Sahni, Acta Metall. 32, 783 (1984).
2. D. J. Srolovitz, M. P. Anderson, P. S. Sahni and G. S. Grest, Acta Metall. 32, 793 (1984).
3. G. S. Grest, M. P. Anderson and D. J. Srolovitz, "Computer Simulation of Microstructural Dynamics", this volume.
4. F. Haessner, "Recrystallization of Metallic Materials" (edited by F. Haessner), Dr. Riederer Verlag GmbH, Stuttgart (1978), p. 2.
5. J. G. Byrne, "Recovery, Recrystallization, and Grain Growth", MacMillan, New York, NY (1965), p. 60.
6. W. A. Johnson and R. F. Mehl, Trans. AIME 135, 416 (1939).
7. M. Avrami, J. Chem. Phys. 7, 1103 (1939); ibid. 8, 212 (1940); ibid 9, 177 (1941).
8. F. Haessner, "Recrystallization of Metallic Materials" (edited by F. Haessner), Dr. Riederer Verlag GmbH, Stuttgart (1978), p. 3.

9. K. Detert, "Recrystallization of Metallic Materials" (edited by F. Haessner), Riederer Verlag, Stuttgart (1978), p. 97.
10. M. Hillert, Acta Metall. 13, 227 (1965).
11. O. Hunderi and N. Ryum, Acta Metall. 29, 1737 (1981).
12. O. Hunderi and N. Ryum, Acta Metall. 30, 739 (1982).
13. J. L. Walter and C. G. Dunn, Acta Metall. 8, 497 (1960).
14. F. H. Buttner, E. R. Funk and H. Udin, J. Phys. Chem. 56, 657 (1952).
15. D. Kohler, J. Appl. Phys. 31, Suppl., 408S (1960).
16. D. J. Srolovitz, G. S. Grest and M. P. Anderson, Acta Metall., submitted.
17. D. J. Srolovitz, G. S. Grest and M. P. Anderson, Acta Metall. 33, 2233 (1985)
18. H. P. Stuwe, "Recrystallization of Metallic Materials" (edited by F. Haessner), Dr. Riederer Verlag GmbH, Stuttgart (1978), p.12.
19. E. N. Gilbert, Ann. Math. Stat. 33, 958 (1962).

INTERFACE MODELS OF DENDRITIC GROWTH

David A. Kessler, Joel Koplik# and Herbert Levine#*

*Dept. of Physics
Univ. of Michigan
Ann Arbor, MI 48109

#Schlumberger-Doll Research
Old Quarry Road
Ridgefield, CT 06877

ABSTRACT

In order to understand the dynamics of pattern selection and sidebranch emission in dendritic growth, we develop two models which describe the time-evolution of the dendrite boundary. In the first model, the motion of the interface is strictly a function of the local geometry of the interface. The model, designed to mimic the physics of the diffusion-controlled growth, is however much simpler to analyze due to its essentially one-dimensional nature. We construct an algorithm to simulate the time development of interfaces in the model. These simulations show that our model grows dendritic-like shapes upon the introduction of sufficient anisotropy, and corroborate our analytic demonstration of velocity selection via a global solvability condition. The second model examined is the infinite diffusion length limit of a more realistic diffusion model. The algorithm developed for the local model can be easily extended to treat this model. Simulations reveal that this model also exhibits dendritic behavior, with a tip-splitting instability for insufficient anisotropy.

1. INTRODUCTION

The characterization of the growth of a single domain is clearly essential to a fundamental understanding of the microstructure of a material. For example, in the dendritic growth regime, the knowledge of growth velocity, tip radius, and sidebranching frequency is a prerequisite to calculations of larger scale microstructure. Despite much study over the last four decades, a theoretical understanding of how to calculate these quantities from first principles is still lacking.

The phenomenon of dendritic growth in solidification has a broader significance outside its obvious metallurgical importance. It represents a striking example of pattern formation in diffusion- controlled systems. Dendritic growth is seen in systems ranging from electrochemical cells[1] to two-phase fluid flow in anisotropic Hele-Shaw cells.[2] A better understanding of the solidification problem will contribute to progress in characterizing when dendritic patterns are expected in general and under what conditions transitions to other growth regimes might occur.

Let us consider, for specificity, what might be considered the minimal realistic model of dendritic growth, with an infinitely sharp interface and equal diffusion constants in the solid and liquid. The equations for this model are

$$D\nabla^2 T = \dot{T} \tag{1.1a}$$

$$T(\vec{x}_I(s)) = T_M - \frac{\gamma(\theta)}{L}\kappa(s) \tag{1.1b}$$

$$c_p D\left[(\hat{n}\cdot\vec{\nabla}T)_l - (\hat{n}\cdot\vec{\nabla}T)_c\right] = -L\vec{v}(s)\cdot\hat{n} \tag{1.1c}$$

Here, T is the rescaled temperature field, which satisfies the diffusion equation. The boundary condition (1.b) reflects the effect of surface tension via the Gibbs-Thomson relation, but ignores the corrections due to kinetic effects. The surface tension, γ, has a dependence on θ, the orientation of the interface, due to the underlying crystal anisotropy. The interface is given by $\vec{x}_I(s)$, s being the arclength, and has curvature $\kappa(s)$. Lastly, (1.c) relates the normal velocity of the interface, $\hat{n}\cdot\vec{v}(s)$, to the production of latent heat, which results in a discontinuity of the temperature gradient going from liquid (l) to crystal (c). The material parameters are given by: L, the latent heat of fusion; c_p, the specific heat (assumed equal in solid and liquid);

and D, the thermal diffusivity. Progress in unraveling the mechanisms responsible for dendritic growth has been very difficult to achieve[3] even in the context of this grossly simplified model. Nor have reliable numerical computations been done for this model.

What makes this problem so difficult is its intrinsic non-locality. The temperature field diffuses information throughout the entire system, so a local change of the solid-liquid interface at one point is reflected at all other points on the interface. Not only is the action of the temperature field non-local in space, it is also non-local in time, so that the interface motion is affected by the entire past history of the system. To overcome this barrier, we have sought to eliminate the temperature field from the problem, and focus attention on the interface itself as the dynamic variable of interest. This is desirable, since, in the end, it is the only the interface which is relevant for the morphological questions we seek to address.

In this paper, we will report on two directions in which we have pursued this idea of interface dynamics. The first[4,5] is an attempt to construct a purely local model of interface dynamics, designed to mimic insofar as possible the physical content of the solidification problem. This, then, represents the ultimate truncation of the non-locality inherent in the original equations. We shall see that one can recover, to a surprising degree, the phenomenology of dendritic growth in this simplest of models.

The second model[6] we will discuss is the infinite diffusion length limit of the system (1.1). In this limit, the diffusion equation is replaced by the Laplace equation. This eliminates the non-locality in time, while retaining the non-local interactions between different parts of the interface at the same instant. This model can then be represented as a model for the interface only, with the temperature field eliminated using a boundary integral formulation. Then, using the techniques developed for the local model, we can study numerically the time-development of this system. The results of this study reinforce the conclusions obtained from the local model.

2. THE LOCAL INTERFACE MODEL

In this section, we describe our local model of interface growth in two dimensions. As stated in the introduction, we wish to construct a model whose only degrees of freedom are those associated with the interface itself. The requirement of locality then leads to the result that that motion of a point on the interface must be a function of the local geometry of the interface at that point. The equation of motion of the interface can then be written

$$\vec{v}_I \cdot \hat{n} = F(\theta, \kappa, \frac{\partial \kappa}{\partial s}, \frac{\partial^2 \kappa}{\partial s^2}) \tag{2.1}$$

In this equation, $\hat{n}$ is the normal to the interface, θ is the angle between $\hat{n}$ and a fixed direction, (which we take to be the y-axis), s is the arclength along the curve, and $\kappa = ds/d\theta$ is the (mean) curvature of the interface.

The choice of a particular form of F is arbitrary, albeit motivated by the physics of solidification. We set

$$F = (\gamma \frac{\partial^2 \kappa}{\partial s^2} + \lambda \kappa + A\kappa^2 - B\kappa^3)(1 + \epsilon \cos m\theta) \tag{2.2}$$

Here, γ plays the role of surface tension, acting to suppress short-wavelength fluctuations. The $\lambda\kappa$ term is responsible for the Mullins-Sekerka instability.[6] The parameter A can be viewed as analogous to the undercooling, while B controls the critical nucleation size, below which a circular interface will shrink. The absence of a constant term reflects the fact that the growth velocity of a circular interface goes to zero as the radius increases. We have modeled the effects of the crystal anisotropy, which enhances growth along the m-fold crystal axes, by the ϵ term. In the following we shall focus on the case m=4, corresponding to an underlying cubic symmetry, the results for other values of m being qualitatively similar. We shall also choose units such that γ and λ are set to unity.

We can immediately derive two results, both of which are analogous to results for the more realistic system (1). First, in the absence of surface tension, the model possesses a continuous family of steady-state needle-crystal solutions, which propagate at a constant, arbitrary, velocity. These are the parallels to the Ivantsov family of parabolic needle crystal solutions[7] of (1). Second, the linear stability analysis for a growing circle of radius R yields the result that a perturbation of the form $\cos k\theta$ has a growth rate

$$\sigma_k = (k^2 - 1)\left[\frac{1}{R^2}(1 + \frac{2A}{R} - \frac{3B}{R^2}) - \frac{k^2}{R^4}\right] \tag{2.3}$$

which is also similar to the results obtained for the realistic model.

The advantage of the local interface model is that all the dynamics is reduced to the interface, giving rise to an effectively one-dimensional system. The key to efficiently simulating the time-development of the model is to exploit this by recasting the equations of motion in an explicitly one-dimensional form. We start by realizing that the two-dimensional interface can be reconstructed from the single function $\theta(s)$. The arclength s varies from 0 to s_T, the total arclength. As it is more convenient to have the domain of θ be time-independent, we introduce the relative arclength, $\alpha \equiv s/s_T$. The variables are then $\theta(\alpha,t)$, $0 \leq \alpha \leq 1$, and s_T. Then, the fundamental equation (2.1) implies that

$$\frac{\partial\theta(\alpha,t)}{\partial t} = -\frac{1}{s_T}\frac{\partial F}{\partial\alpha} - s_T\kappa(\alpha)\left[\int_0^\alpha \kappa F d\alpha' - \alpha\int_0^1 \kappa F d\alpha'\right]$$

$$\frac{ds_T}{dt} = s_T\int_0^1 \kappa F d\alpha \tag{2.4}$$

where $\kappa = \frac{1}{s_T}\frac{\partial\theta}{\partial\alpha}$. The equation for s_T accounts for the stretching of a curved piece of interface due to the fact that the different ends are moving away from each other. The equation for θ has two terms. The first reflects the fact that when the velocity varies along a point of the interface, the angular orientation of this segment rotates. The second term arises from the non-local adjustment needed to stay at a fixed α as the entire curve stretches and contracts in various places. The problem has now been reduced to solving the first-order system (2.4), whereupon the interface at time t is given by

$$x(\alpha,t) = s_T\int_0^\alpha \cos\theta(\alpha',t)\, d\alpha'$$

$$y(\alpha,t) = s_T\int_0^\alpha \sin\theta(\alpha',t)\, d\alpha' \tag{2.5}$$

There are three constraints on θ which must be satisfied. The first expresses the periodicity of θ: $\theta(\alpha=1) = \theta(\alpha=0)$. The other two enforce the closing of the curve: $x(\alpha=1) = x(\alpha=0)$; and $y(\alpha=1) = y(\alpha=0)$. The equations of motion (3.1) are, of course, consistent with all three constraints; i.e., if θ satisfies the constraints at $t=0$ then θ will continue to satisfy them for all time In a numerical solution of (2.4), however, care must be exercised to maintain these constraints. The first is trivial to enforce, but the last two, being nonlinear in θ, present a greater challenge. If the curve has an m-fold ($m \geq 2$) rotational symmetry, however, these constraints are also trivially satisfied. We choose then to only consider initial conditions which possess the same $m(=4)$ -fold symmetry as the anisotropy. For convenience, we also restrict ourselves to initial conditions which are reflection symmetric about each m-fold symmetry axis. As both these symmetries are preserved by the equations of motion, we then need only solve (2.4) for $0 \leq \alpha \leq 1/2m$ and use the symmetries to extend the solution to the full range of α. This greatly reduces the computational burden. A stable method of handling the non-linear constraints in the non-symmetric case would be very useful for future work.

The next step is to discretize (2.4). We let $\theta_j \equiv \theta(\alpha_j)$ where $\alpha_j = \frac{j}{2mN}$, $j=0,\ldots,N$ and $\theta_0 \equiv 0$, $\theta_N \equiv \frac{\pi}{m}$. Replacing integrals and derivatives by finite sums and differences, (2.4) is transformed to a first-order set of N coupled ordinary differential equations. We then simply give this system to the packaged routine LSODE,[8] which uses a predictor-corrector method adapted to handle stiff systems, to integrate forward in time. The input to this routine includes an evaluation of the Jacobian matrix $M_{i,j} \equiv \frac{\partial \dot{\theta}_i}{\partial \theta_j}$, which we compute analytically from the discretized form of (2.4). Alternatively, at a minor cost in speed and accuracy, LSODE will itself compute $M_{i,j}$ numerically. The computation time scales as N^3 and a typical run of 300 points takes 8-10 hours on the VAX 11/780 in double precision.

Several aspects of our algorithm should be noted. We have opted for a parametrization in terms of relative arclength. This has the effect of introducing a degree of nonlocality into the equations. In consequence, the Jacobian matrix M is not banded or sparse and the required inversion of M is slow. An alternative formulation involving a time-independent parametrization would not have this problem. The cost, however, is that the arclength differences between the discretized points then vary in time so that derivatives become less accurate and dynamical reassignment of points is required. This reassignment is an additional source of noise, and furthermore would result in reduced efficiency of the predictor-corrector algorithm.

We first examine the result of a simulation with no anisotropy; i.e., $\epsilon = 0$. A typical example, with A=4, B=-.25, is shown in Fig. 1. We started with a perturbed circle for the initial interface, given by $\theta_j = \frac{\pi j}{mN} + \frac{\delta}{m}\sin\frac{\pi j}{N}$, $s_T = 2\pi r_0$, $\delta = .1, r_0 = 10$. Thus, the interface initially has tips. These tips grow for a short time and then widen and split into two. The new tips thus produced also grow out for a period. In the final stage shown, they are widening and about to split again. This sequence of repeated tip-splitting is generic for the isotropic model, independent of A and B and initial conditions.

The behavior of the model in the presence of anisotropy is markedly different. In Figure 2, we display the results of a simulation with A=4, B=1.5, $\epsilon = .15$, and initial conditions parametrized by $\delta = 3.6$, $r_0 = 8$. Here, the initial tips persist and move outward while emitting several sidebranches. In addition, the sidebranches themselves grow stably outward and develop their own secondary sidebranches.

The effect of anisotropy can be best illustrated by examining the velocity of the tips as a function of time. This is plotted in Figure 3 for the cases ϵ = .085, .125, .15, and .2, with A=4, B=1.5 fixed. The oscillatory behavior is correlated with the sidebranching, with each minimum corresponding to the emission of a new sidebranch. At small values of ϵ, these oscillations are unstable and eventually cause the tips to split. At the critical anisotropy, $\epsilon_c \approx .15$ for this set of parameters, stable oscillations occur and sidebranching persists indefinitely. For larger ϵ, the oscillations are damped and the sidebranching dies out, as the tip becomes an unadorned needle crystal. It is noteworthy that all three modes of behavior; i.e., tip-splitting, persistent sidebranching, and pure needle crystal growth have all been

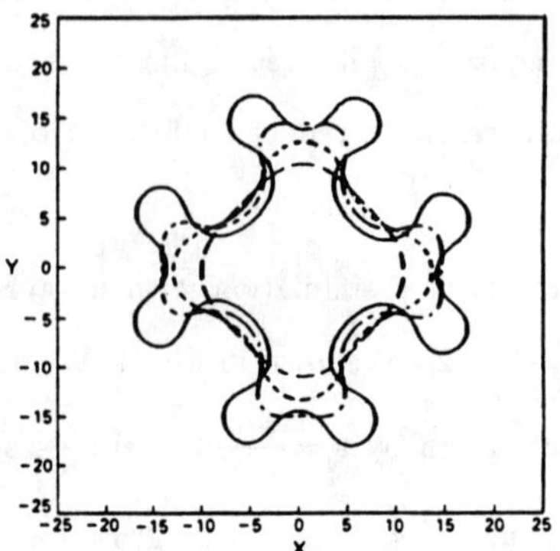

Figure 1. Tip-splitting sequence in local model (A=4, B=.25, eps=0)

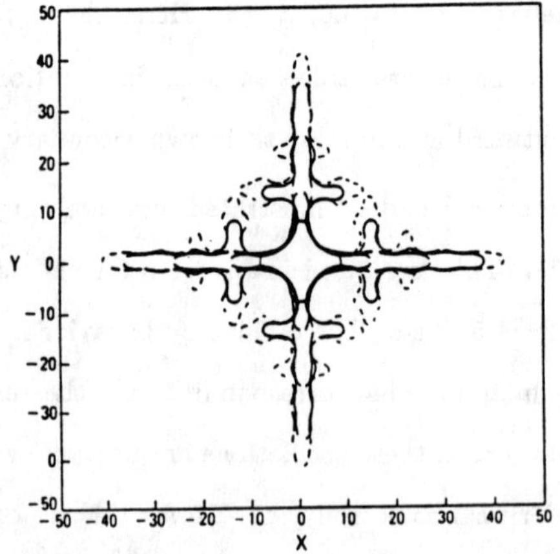

Figure 2. Sidebranching sequence in local model (A=4, B=1.5, eps=.15)

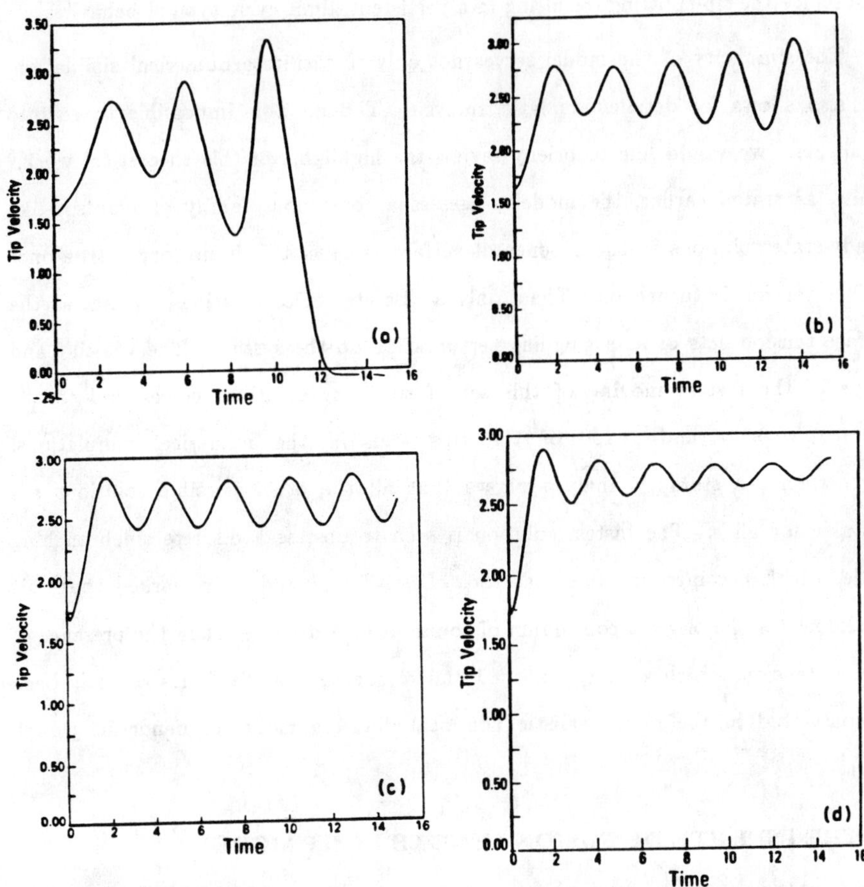

Figure 3. Tip velocity vs. time for various anisotropies in local model (A=4, B=1.5): a) eps=.085; b) eps=.125; c) eps=.15; d) eps=.2

seen in real materials. The robustness of the dendritic (persistent sidebranching) morphology in real systems is a puzzle, however, as it occurs in our model at only a single value of ϵ. In real systems, it is possible that non-linear effects act, in general, to stabilize the tip-splitting, resulting in a persistent, limit-cycle type of behavior.

The simplicity of the model serves not only to facilitate numerical simulation, but also allows for detailed analytic analysis. Though it is impossible to go into detail here, we would like to briefly review the highlights of this theoretical work.[5] While, as stated earlier, the model possesses a continuous family of Ivantsov-like steady-state solutions in the absence of surface tension, this is no longer true once surface tension is turned on. Then, only a discrete set of solutions remain, as the surface tension acts of as a singular perturbation on the system. The velocity and shape of the fastest moving of this set of surviving solutions correspond to the average velocity and shape of the tips seen in the numerical simulations. Furthermore, a stability analysis shows that all the other possible solutions are unstable for all ϵ. The fastest solution is seen to possess a discrete sidebranching mode, which becomes unstable for $\epsilon<\epsilon_c$. Lastly, it should be remarked that this breakdown of the original continuum of solutions to a discrete set in the presence of surface tension, which we have labelled "microscopic solvability" has recently been demonstrated in the more realistic non-local diffusion model of dendritic growth (1.1).[9]

3. THE INFINITE DIFFUSION LENGTH LIMIT MODEL

In this section, we turn to our second interfacial model for dendritic growth. This is the infinite diffusion length limit of the model describe in Eq. (1.1). In this limit, the diffusion equation for the temperature field simplifies to the Laplace equation, and the motion of the interface at a given time in solely a function of the shape of the interface at that time. We can see this explicitly by eliminating the temperature field altogether via a boundary integral method. After appropriate rescalings, the equation of motion can be recast in the form

$$1 - \kappa(s)\,(1 - \epsilon\cos m\theta) = \int_0^{s_T} G(\vec{x}(s),\vec{x}(s'))\,ds' \tag{3.1}$$

where G is the Green's function for the Laplace operator

$$G(\overline{x},\overline{y}) = -\ln(\overline{x}-\overline{y})^2 + \ln(\frac{R^2\overline{y}}{y^2}-\overline{x})^2 + \ln\frac{R^2}{y^2} \tag{3.2}$$

Here we have assumed a cold bath at a far distance R, and taken an angular dependence for the surface tension of the form $\kappa(1-\epsilon\cos(\theta))$. The latter is obviously a gross simplification, but the qualitative results are independent of the precise form. Now, given the interface at time t, equation (3.1) implicitly determines the velocity of every point on the interface at that time. Given the velocity, we can use the analysis of the last section and step the interface forward in time by means of the $\theta(\alpha)$ representation. Thus, we proceed by first generating the real space coordinates of the curve from θ. Then, by discretizing the integral in (3.1), we obtain a matrix equation for $v(\alpha)$. In the discretization, care must be taken to handle the logarithmic singularity in the Green's function at s=s'. We solve the matrix equation using a standard routine, and then continue as in section 2, with F replaced by v. The computation time again scales like N^3, though the bottleneck is now the calculation of the Jacobian matrix M, rather than its inversion, as previously. A typical run with N=100 takes 2 hours on the FPS-164 array processor.

The results of these simulations are displayed in Fig. 4. In Fig. 4a, we see the tip-splitting at zero anisotropy. The results, for ϵ=.2, shown in Fig. 4b, demonstrate how finite anisotropy has stabilized this instability, resulting in a dendritic morphology. This, of course, is precisely parallel to what we observed in our local interface model.

We have thus seen how, by focusing on the dynamics of the interface itself, we have succeeded in furthering our understanding of the dendritic growth process. We feel that this approach has much to offer in the investigation of other systems featuring a well-defined interface of non-trivial morphology.

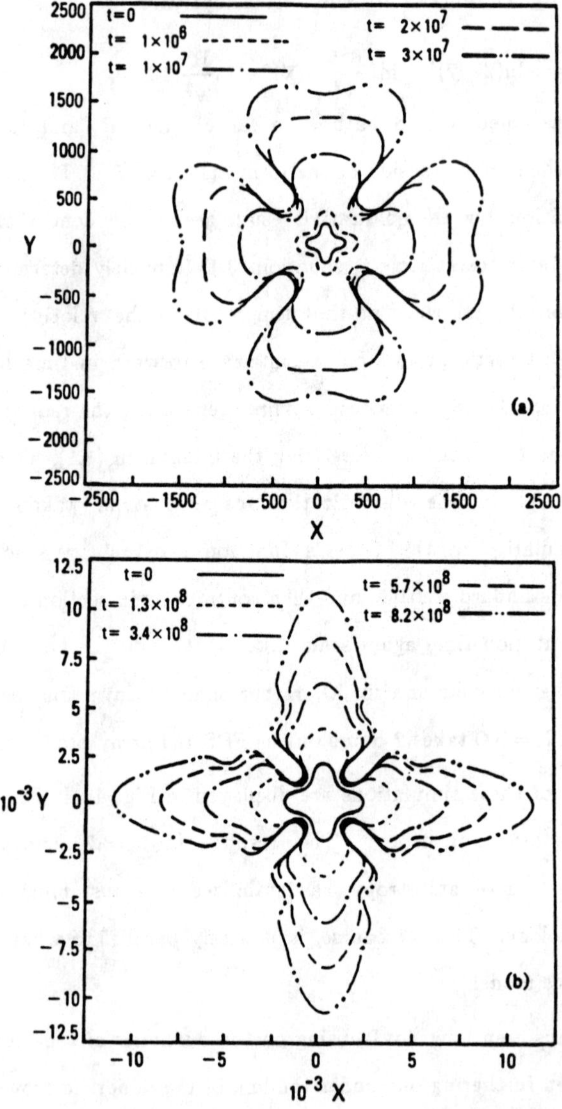

Figure 4. Time-development of interface in infinite diffusion length limit (R=200): a) eps=0 b) eps=.2

REFERENCES

1 Y. Sawada, A. Dougherty, and J. Gollub, Haverford preprint (1985); D. Grier, E. Ben-Jacob, R. Clarke, and L. M. Sander, Michigan preprint (1985).

2 E. Ben-Jacob, R. Godbey, N. D. Goldenfeld, J. Koplik, H. Levine, T. Mueller, and L. M. Sander, Phys. Rev. Lett. 55, 1315 (1985).

3 For a review of much of the earlier work on the problem of dendritic growth, see J. S. Langer, Rev. Mod. Phys. 52, 1 (1980).

4 R. C. Brower, D. Kessler, J. Koplik, and H. Levine, Phys. Rev. Lett. 51, 1111 (1983); Phys. Rev. A29, 1335 (1984); D. Kessler, J. Koplik, and H. Levine, Phys. Rev. A30, 3161 (1984).

5 D. Kessler, J. Koplik, and H. Levine, Phys. Rev. A31, 1712 (1985).

6 W. W. Mullins and R. F. Sekerka, J. Appl. Phys. 34, 323 (1964).

7 G. P. Ivantsov, Dokl. Akad. Nauk. USSR 58, 567 (1947).

8 A. C. Hindmarsh, ACM-Signum Newsletter 15, 10 (1980)

9 D. Kessler and H. Levine, Schlumberger-Doll Preprint (1985); D. Meiron, CalTech preprint (1985).

MICROSTRUCTURAL COARSENING IN 2- AND 3-DIMENSIONS—APPLICATIONS OF MULTIPARTICLE DIFFUSION ALGORITHMS

M. E. Glicksman and S. P. Marsh

Materials Engineering Department
Rensselaer Polytechnic Institute
Troy, NY 12180-3590

Abstract

Microstructural evolution can occur via diffusional interactions among isolated particles or domains embedded in a continuous matrix phase. In 3-dimensions, scaling arguments indicate kinetic behavior similar to that suggested by Lifshitz, Slyozov, and Wagner (LSW). Application of 3-dimensional multiparticle algorithms by Voorhees and Glicksman has shown that linear features grow as the cube-root of time for a range of particle volume fractions. Chakraverty has treated the case of diffusion on a 2-dimensional surface in a manner analogous to LSW, and found that linear features grow as the fourth-root of time. Again, we have developed a 2-dimensional multiparticle algorithm which shows significant departures from Chakraverty's results occur at low and high area coverages. A brief discussion of computational techniques will also be provided.

Introduction

The properties and behavior of many inhomogeneous materials are affected by the simultaneous diffusional growth and dissolution of a dispersed second-phase embedded in a continuous matrix. Three-dimensional multiparticle diffusion, or "Ostwald ripening", occurs in such diverse metallurgical phenomena as the coarsening of fine primary-phase structures in solid-liquid mixtures, sintering, the aging of precipitates from supersaturated solid solutions, and void growth during fast neutron irradiation. Diffusion on a two-dimensional surface or substrate can occur in such processes as the sintering of supported metal catalysts and the aging of discontinuous thin films.

The driving forces for these processes vary, but they generally involve gradients of temperature or chemical concentration caused by the radius-dependent solubility of the dispersed second-phase particles. A solution to the diffusion equation for the entire two-phase mixture is needed to model accurately the kinetics of such processes. Many previous analyses of multiparticle diffusion problems (MDP's) rely on unrealistic assumptions to obtain approximations to the diffusion field. In this work we present a theoretical approach applied to both 2 and 3-dimensional MDP's which avoids many of these limiting assumptions by modeling the diffusion field as a dimensionless potential field. Lattice potential summation techniques, adapted from a method first used by Ewald (1), are used to obtain a detailed solution to the diffusion field which includes specific interparticle effects.

Background

A major advance in the analysis of the 3-dimensional MDP is the continuum statistical approach developed by Lifshitz and Slyozov (2) and Wagner (3). In the LSW method, each individual spherical particle is assumed to interact with a "mean field" located an infinite distance away. The diffusion field around the particle is determined analytically, and the flux of heat or solute entering or leaving the particle is calculated as a function of the particle size and "mean-field" value only. The "mean-field" value is set by the average particle size of an infinite ensemble of particles, which is assumed to form a continuous size distribution. The hydrodynamic continuity equation is then applied to the particle size distribution (PSD) using the derived single-particle flux functions.

The relevant conclusions of that analysis are as follows:

1. The PSD becomes an asymptotic, invariant distribution when the particle radii are scaled by the average radius of the distribution.
2. The maximum radius of a particle in the asymptotic PSD always remains 1.5 times larger than the average radius.

3. The cube of the average radius (or first moment) of the PSD grows linearly with time at a scaled rate of 4/9.

A major drawback to the LSW statistical continuum approach is that it assumes a vanishingly small volume fraction of the dispersed second phase. This assumption allows individual particle fluxes to be calculated analytically but is physically unrealistic in observed systems. A more detailed description of the LSW analysis is described elsewhere (4), as are some attempts to modify the LSW approach to account for finite volume-fraction effects, none of which are completely satisfactory.

Another approach to solving the MDP is a discrete direct summation technique, where the individual particle diffusion fields are superposed to obtain the overall field. The first significant application of this method is that of Weins and Cahn (5). Although it does account for specific particle interactions, their approach was limited to treating just a few particles. Therefore, the results cannot be reliably extended to provide statistical information on the coarsening kinetics of a large uniform dispersion of second-phase particles.

The continuum statistical approach to the 2-dimensional MDP was developed by Chakraverty (6), in a manner analogous to the LSW method. The particles were modeled as hemispheres situated on a flat substrate. In Chakraverty's analysis, however, the radially symmetric diffusion field for each individual particle extends to a "screening distance" equal to e (2.718...) times the particle radius, where the concentration is assumed to reach its uniform average value. This "screening distance" is infinite in the LSW approach (indicating a vanishingly small volume fraction of second-phase material), but must be assumed finite here because the single-particle potential in two dimensions is logarithmic in character and therefore becomes divergent both at large and small distances. Using the finite screening distance assumption, Chakraverty determined the flux entering or leaving a particle in terms of the particle size. He applied the hydrodynamic continuity equation to the 2-dimensional PSD using this flux function, and concluded the following:

1. The PSD approaches a time-independent distribution when the particle radii are scaled by the average radius of the ensemble (as in the LSW case).
2. The maximum radius of a particle in this distribution is 4/3 as large as the average radius.
3. The fourth power of the average radius grows linearly with time, at a scaled rate of 27/64.

Chakraverty's analysis is outlined in more detail elsewhere (7). The main drawback in that theory is the arbitrary "screening distance" imposed on each particle, since the calculated particle fluxes are sensitive to this distance. Again, specific particle interaction effects were ignored, as were the effects of distinct fractional surface coverages on diffusion rates.

Analysis

An alternative description and approximate solution of the multiparticle diffusion field will now be developed. Physically, any collection of spherical particles in a three-dimensional matrix, or a collection of hemispheres on a flat substrate, will continuously absorb and emit heat or solute atoms at their boundaries. In such a system, where the matrix serves only to transmit the fluxes of heat or mass, the larger particles tend to grow while the smaller ones tend to shrink. This behavior results from the application of the Gibbs-Thomson boundary condition to each particle. For the case of mass transfer, this condition states that smaller particles are in equilibrium with a relatively higher solute atom concentration at their boundaries because of their high curvature. Mass tends to diffuse away from these smaller particles, causing them to shrink even more. Solute atoms tend to diffuse toward larger particles, because their equilibrium perimetric concentration is relatively lower. Similar arguments hold for the case of heat transfer, with smaller particles having a relatively lower equilibrium temperature at their highly-curved boundaries than do larger particles. The resulting temperature gradients in the matrix tend to diffuse heat toward the smaller particles, causing them to melt, while larger particles tend to grow.

It can be seen from this discussion that each particle affects the growth rate of nearby particles by influencing the local availability of heat or solute. Each particle is thus a source or sink of heat or solute, depending on whether it is growing or shrinking at a given moment. In potential theoretic terms, each particle can be modeled as an equivalent point source of strength B, where B is proportional to the volumetric growth rate of a particle. B is chosen to be negative for a shrinking particle, and will be a function of time for a given particle as it and the surrounding particles change in size.

If the ripening process is diffusion-controlled, so that interfacial absorption/desorption is not the rate-limiting step, then the quasi-static diffusion field can be approximated by Laplace's equation. The variables in the diffusion equation can be appropriately scaled for either heat or mass diffusion (7,8), so that θ represents either a dimensionless concentration or temperature. The resulting form of Poisson's equation which describes the diffusion field arising from a collection of N point sources/sinks is as follows:

$$\nabla^2 \theta = \sum_{j=1}^{N} -2\pi B_j \, \delta(\vec{r} - \vec{r}_j) \quad [2\text{-D}] \quad , \tag{1a}$$

and $$\nabla^2\theta = \sum_{j=1}^{N} -4\pi B_j \, \delta(\vec{r} - \vec{r}_j) \quad [3\text{-D}] \tag{1b}$$

In these equations B_j is the source strength of the jth particle, $\vec{r}_j$ is the location of the center of the jth particle, $\vec{r}$ is an arbitrary field point, and δ is the Dirac delta function.

The source strength B_j represents the dimensionless volumetric growth rate of the jth particle. For a system of N particles which exchange mass or heat through the matrix, where there are no external sources or sinks (no net precipitation), the mass/heat conservation equation can be written as

$$\sum_{j=1}^{N} B_j = 0. \tag{2}$$

This equation is equivalent to the statement that the total volume of second-phase particles remains constant with time.

Based on the linearity of Poisson's equation, a solution to Equations 1a and 1b may be written as the linear combinations

$$\theta(\vec{r}) = B_o + \sum_{j=1}^{N} B_j \log|\vec{r} - \vec{r}_j| \quad [2\text{-D}], \tag{3a}$$

and $$\theta(\vec{r}) = B_o + \sum_{j=1}^{N} \frac{B_j}{|\vec{r} - \vec{r}_j|} \quad [3\text{-D}], \tag{3b}$$

where B_o is a constant reference potential. The multiparticle diffusion field can be calculated straightforwardly from Equations 3a and 3b, but two difficulties arise when applying this method to an arbitrarily large number of particles. The first is that the summation in Equation 3a is divergent for large N because of the logarithmic terms, whereas the summation in Equation 3b is semi-convergent, and the limit can depend on the order in which the terms are summed. A second problem encountered when attempting direct summations is that a scheme is needed to specify the locations of all the dispersed particles.

These difficulties can be averted by creating a finite "basis" of random particles and selecting a relatively large lattice vector, $\vec{r}_e$. The basis cell will be a square for 2-D diffusion, and a cube for the 3-D case. The basis is filled with a finite number of randomly located particles arranged in the desired size distribution. The size (repeat distance) of the basis is then determined by the desired surface coverage (2-D) or volume fraction (3-D). The basis is then repeated by translation indefinitely in two or three dimensions to fill all space in a Bravais lattice configuration. Each allowed lattice translation vector, $\vec{r}_e$, locates the origin of a translated image of the basis in the Bravais lattice. A two-dimensional diagram of this configuration is pictured in Figure 1, where the basis and eight surrounding translations of the type <10> and <11> are shown. In this figure, the basis contains 29 particles.

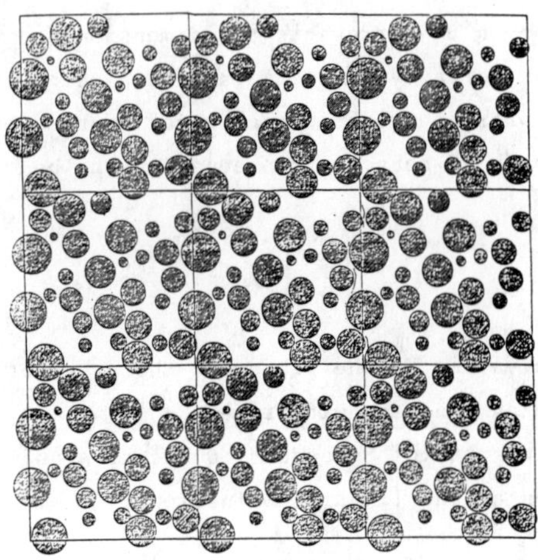

Figure 1 - System of particles consisting of a basis and eight surrounding translated cells.

The solution to Laplace's equation in two and three dimensions, given by Equations 3a and 3b, can be reformulated in terms of the basis and lattice translation vectors as

$$\theta(\vec{r}) = B_o + \sum_{r_e} \sum_{i=1}^{N'} B_i \log |\vec{r} - \vec{r}_i - \vec{r}_e| \quad [2\text{-D}], \qquad (4a)$$

and

$$\theta(\vec{r}) = B_o + \sum_{r_e} \sum_{i=1}^{N'} \frac{B_i}{|\vec{r} - \vec{r}_i - \vec{r}_e|} \quad [3\text{-D}], \qquad (4b)$$

where N' is the number of particles in the basis. The advantage to this reformulation is two-fold. First, a uniform spatial distribution containing an infinite number of particles is completely specified by the locations of only N' particles and the size of the basis. Secondly, Equations 4a and 4b now resemble lattice potential equations, for which there exist mathematical transformations which convert the lattice sums into rapidly and absolutely convergent summations. This technique was first developed by Ewald (1) to determine the coulombic potential in alkali halide structures. Although the source strengths in an ionic lattice are known and constant, whereas the B_i's in a diffusion problem are as yet undetermined and vary with time, the Ewald technique can still be applied to Equations 4a and 4b to obtain a rapidly convergent form of the dimensionless potential field $\theta(\vec{r})$. The full adaptation of Ewald's method to the MDP is described in detail elsewhere (4,7), and results in the following solutions:

$$\theta(\vec{r}) = B_o + \left(\frac{2\pi}{a_o^2}\right)\sum_{i=1}^{N'}\sum_{\vec{g}}{}' \frac{B_i}{G^2}\exp\left[\frac{-G^2 a_o^2}{4\pi} + i\,\vec{g}\cdot(\vec{r}-\vec{r}_i)\right] + \frac{1}{2}\sum_{i=1}^{N'}\sum_{\vec{r}e} B_i\,\mathrm{Ei}\left[-\frac{\pi}{a_o^2}\left|\vec{r}-\vec{r}_i-\vec{r}_e\right|^2\right] \quad [2\text{-D}] \quad , \qquad (5a)$$

$$\theta(\vec{r}) = B_o + \left(\frac{4\pi}{a_o^3}\right)\sum_{i=1}^{N'}\sum_{\vec{r}_e} \frac{B_i}{G^2}\exp\left[\frac{-G^2 a_o^2}{4\pi} + i\,\vec{g}\cdot(\vec{r}-\vec{r}_i)\right] + \sum_{i=1}^{N'}\sum_{\vec{r}_e} \frac{B_i\,\mathrm{erfc}\left[\left(\frac{\pi}{a_o^2}\right)^{1/2}\left|\vec{r}-\vec{r}_i-\vec{r}_e\right|\right]}{\left|\vec{r}-\vec{r}_i-\vec{r}_e\right|} \quad [3\text{-D}] \ . \qquad (5b)$$

The above equations allow the dimensionless potential θ to be calculated from the $N' + 1$ sources, B_i (including B_o), and the locations of the N' particles in the basis. The B_i values, which are proportional to the growth rates of the particles, are as yet undetermined. Their values can be found by applying the Gibbs-Thomson boundary condition to each particle. The linearized Gibbs- Thomson equation can be written in terms of dimensionless variables as

$$\theta(R_i) = \frac{1}{R_i} \text{ (mass transfer)} \quad , \qquad (6a)$$

or

$$\theta(R_i) = -\frac{1}{R_i} \text{ (heat transfer)} \quad , \qquad (6b)$$

for both the 2- and 3-dimensional cases considered, where R_i is the dimensionless radius of the ith particle.

This condition can be strictly applied in the case of a single-particle diffusion field, which is radially or spherically symmetric. In the more complex multiparticle diffusion field, the potential θ will vary along the perimeter of a hemispherical particle (2-D) or over the surface of a sphere (3-D). This boundary condition can be applied, however, if it is set equal to the perimeter- or surface-averaged potential, $\bar{\theta}$. In both two and three dimensions, this reduces to setting the potential at the center of the ith particle, $\theta(\vec{r}_i)$, equal to $\frac{1}{R_i}$. Applying this averaged boundary condition to the ith particle and using Equations 5a and 5b for $\theta(\vec{r})$ leads to the following relations between B_i and R:

$$\frac{1}{R_i} = B_o + B_i \log R_i + \left(\frac{2\pi}{a_o^2}\right)\sum_{j=1}^{N'}\sum_{\vec{g}}{}' \frac{B_j}{G^2} \exp\left[\frac{-G^2 a_o^2}{4\pi} + i\,\vec{g}\cdot(\vec{r}_j - \vec{r}_i)\right]$$
$$+ \left(\frac{1}{2}\right)\sum_{\substack{j=1 \\ (j\neq i)}}^{N'}\sum_{\vec{r}} B_j\, Ei\left[-\frac{\pi}{a_o^2}\left|\vec{r}_i - \vec{r}_j - \vec{r}_e\right|^2\right]$$
$$+ \left(\frac{B_i}{2}\right)\left(\gamma + \log\left[\frac{\pi}{a_o^2}\right] + \sum_{\vec{r}_e}{}' Ei\left[-\frac{\pi}{a_o^2}\left|\vec{r}_e\right|^2\right]\right) \quad [2\text{-D}] \qquad (7a)$$

$$\frac{1}{R_i} = B_o + B_i\left[\frac{1}{R_i} - \frac{2}{a_o}\right] + \left(\frac{4\pi}{a_o^3}\right)\sum_{j=1}^{N'}\sum_{\vec{g}}{}' \frac{B_j}{G^2} \exp\left[\frac{-G^2 a_o^2}{4\pi} + i\,\vec{g}\cdot(\vec{r}_j - \vec{r}_i)\right]$$
$$+ \sum_{\substack{j=1 \\ \left(\vec{r}_e=0 \atop i\neq j\right)}}^{N'}\sum_{\vec{r}_e} \frac{B_i\, \mathrm{erfc}\left[\left|\vec{r}_j - \vec{r}_i - \vec{r}_e\right| \left(\frac{\pi}{a_o^2}\right)^{\frac{1}{2}}\right]}{\left|\vec{r}_j - \vec{r}_i - \vec{r}_e\right|} \quad [3\text{-D}]\ . \qquad (7b)$$

In Equation 7a, γ represents Euler's constant.

There exist N' independent equations of the type shown in Equations 7a and 7b, which are obtained by applying the averaged Gibbs-Thomson boundary condition to each particle in the basis. However, an additional equation is needed to solve for the $N' + 1$ unknowns, $B_o \ldots B_{N'}$. This additional equation is a modification of the mass/heat conservation condition of Equation 2. Inasmuch as all translated images of the basis behave identically, the conservation of mass or heat can be restated as a local requirement, viz.,

$$\sum_{i=1}^{N'} B_i = 0\ , \qquad (8)$$

where the summation is carried out over the particles in the basis.

There are now $N' + 1$ unknowns, and $N' + 1$ independent equations that relat them. These equations are the N' boundary conditions of the type in Equations 7a and 7b, and the local conservation condition (Equation 8). To determine the source strengths B_i, these equations can be manipulated to form the linear system

$$\{Y\} = [\chi]\ \{B\}\ , \tag{9}$$

where

$$\{Y_i\} = \begin{bmatrix} \pm\ \frac{1}{R_1} \\ \pm\ \frac{1}{R_2} \\ \cdot \\ \cdot \\ \cdot \\ \cdot \\ \cdot \\ \pm\ \frac{1}{R_{N'}} \\ 0 \end{bmatrix} \quad \text{and} \quad \{B\} = \begin{bmatrix} B_1 \\ B_2 \\ \cdot \\ \cdot \\ \cdot \\ \cdot \\ \cdot \\ B_{N'} \\ B_o \end{bmatrix} \tag{10}$$

The symmetric coefficient matrix $\{\chi\}$ can be written as follows:

$$[\chi] = \begin{bmatrix} D_{11} & A_{12} & A_{13} & \text{-------} & 1 \\ A_{21} & D_{22} & A_{23} & \text{-------} & 1 \\ A_{31} & A_{32} & D_{33} & \text{-------} & 1 \\ \cdot & \cdot & \cdot & \cdot\ \cdot\ \cdot\ \cdot & \\ \cdot & \cdot & \cdot & \cdot\ \cdot\ \cdot\ \cdot & \\ \cdot & \cdot & \cdot & \cdot\ \cdot\ \cdot\ \cdot & \\ 1 & 1 & 1 & \text{--------} & 0 \end{bmatrix} \tag{11}$$

The terms appearing on the main diagonal are defined as

$$D_{ii} = \log R_i + \left(\frac{2\pi}{a_o^2}\right)\sum_{\vec{g}}{}' \frac{1}{G^2} \exp\left[\frac{-G^2 a_o^2}{4\pi}\right]$$

$$+ \left(\frac{1}{2}\right)\sum_{\vec{r}_e}{}' \mathrm{Ei}\left[-\frac{\pi}{a_o^2}\,|\vec{r}_e|^2\right] + \frac{1}{2}\left(\gamma + \log\frac{\pi}{a_o^2}\right) \quad [2\text{-D}]\ , \quad (12a)$$

$$D_{ii} = \frac{-2}{a_o} + \frac{4\pi}{a_o^2}\left[\sum_{\vec{g}}{}' \frac{1}{G^2} \exp\frac{-G^2 a_o^2}{4\pi}\right] + \sum_{\vec{r}_e}{}' \frac{\mathrm{erfc}\left[|\vec{r}_e|\left(\frac{\pi}{a_o^2}\right)^{\frac{1}{2}}\right]}{|\vec{r}_e|}$$

$$[3\text{-D}]\ , \quad (12b)$$

and the off-diagonal terms A_{ij} are defined as

$$A_{ij} = A_{ji} = \left(\frac{2\pi}{a_o^2}\right)\sum_{\vec{g}}{}' \frac{1}{G^2} \exp\left[\frac{-G^2 a_o^2}{4\pi} + i\,\vec{g}\cdot(\vec{r}_j - \vec{r}_i)\right]$$

$$+ \left(\frac{1}{2}\right)\sum_{\vec{r}_e} \mathrm{Ei}\left[-\frac{\pi}{a_o^2}\,|\vec{r}_j - \vec{r}_i - \vec{r}_e|^2\right] \quad [2D] \quad (13a)$$

$$A_{ij} = A_{ji} = \left(\frac{4\pi}{a_o^3}\right)\left[\sum_{\vec{g}}{}' \frac{1}{G^2} \exp\left[\frac{-G^2 a_o^2}{4\pi} + i\,\vec{g}\cdot(\vec{r}_j - \vec{r}_i)\right]\right.$$

$$\left. + \sum_{\vec{r}_e} \frac{\mathrm{erfc}\left[|\vec{r}_j - \vec{r}_i - \vec{r}_e|\left(\frac{\pi}{a_o^2}\right)^{\frac{1}{2}}\right]}{|\vec{r}_j - \vec{r}_i - \vec{r}_e|}\right] \quad [3\text{-D}] \quad (13b)$$

These matrix elements, D_{ii} and A_{ij}, contain rapidly convergent sums which result from the application of the Ewald transformation to the boundary conditions of the MDP. The coefficient matrix $[\chi]$ may be termed the particle interaction matrix, since it relates the source strength B_i of each particle in the basis to the source strength of every other particle in the basis and all particle images in the translated cells through the averaged Gibbs-Thomson boundary conditions. The net result of this analysis is that Equation 9, with the definitions given in Equations 10 - 13, provides a unique determination of the diffusion-controlled growth rate of each particle in the basis based on their instantaneous radii and locations. These growth rates are calculated directly from the 2- and 3-dimensional multiparticle diffusion fields, which are found here without recourse to the limiting assumptions of either low volume fraction, a finite number of particles, or of a fixed "screening distance".

Ripening Simulations

In order to obtain information on the kinetics of diffusion-controlled multiparticle ripening, the source strengths B_i must be explicitly related to the particle growth rates. Based on the non-dimensionalizing parameters and the definition of B_i in Equation 1, it can be shown that for both the 2-D and 3-D cases that

$$\dot{R}_i \equiv \frac{B_i}{R_i^2} , \tag{14}$$

where $\dot{R}_i \equiv dR_i/dt$. The ripening behavior may then simulated by solving Equation 9 for the instantaneous B_i values and choosing a small time interval Δt for evolution of the system. The evolving particle sizes are found numerically by using a linearization of Equation 14 of the form

$$R_i(t+\Delta t) \cong R_i(t) + \frac{B_i(t)}{R_i^2(t)} \Delta t. \tag{15}$$

This time-stepping procedure can be continued by alternately applying Equation 9 to find the B_i values at the beginning of each time interval and then applying Equation 15 to calculate the new particle radii at the end of each time interval, Δt. A flowchart for this algorithm is shown in Figure 2. By using this discrete simulation technique the particle size distribution in the basis will evolve with time, with the larger particles tending to grow, while the smaller ones shrink. The time intervals Δt are chosen small enough so that the linearization of Equation 15 accurately describes the kinetic behavior over each interval.

The use of Laplace's equation (Equation 1) to describe the multiparticle diffusion field requires that the supersaturation be small. This condition is met when the scaled particle radii are large ($R_i \gg 1$). To ensure that this condition is maintained during numerical simulations, a particle is assumed to have completely dissolved when it shrinks below some minimum radius and is removed from the distribution. This is accomplished mathematically by eliminating the row and column corresponding to the dissolved particle from the interaction matrix $[\chi]$, thus reducing by one the number of simultaneous equations. All other elements in the matrix remain unchanged, because the locations of the other particles in the basis remain fixed by assumption.

The number of particles initially placed in the basis, N', should be as large as practicable to provide maximum spatial randomness and yield a statistically smooth size distribution. This number is limited by the computational time required to solve the $(N' + 1)$ simultaneous equations for the B_i values at each time-step.

Using the above guidelines, computer simulations of multiparticle ripening

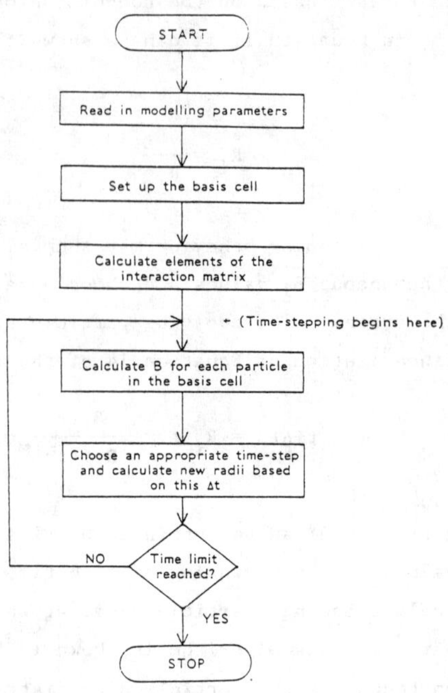

Figure 2 - Flowchart for the numerical ripening simulations.

have been carried out in both two and three dimensions under a variety of initial conditions. The 2-D multiparticle diffusion model has been used to obtain values of the flux B_i as a function of particle size for various values of the surface coverage. These B_i values are multiplied by the average particle radius of the ensemble, $\bar{R}$, to form a modified flux function $\bar{B}$ that is <u>independent</u> of the size scale of the system. This definition is consistent with Chakraverty's scaling arguments in 2-D, which indicate that $\bar{R}^4$ will grow linearly with time. The resulting $\bar{B}$ values from the 2-D ripening simulations show noticeable deviations from the analytic function derived by Chakraverty, who assumed that each particle interacts with a uniform mean field located at a "screening distance" equal to about three times the particle radius. The numerical simulation data shows good agreement with a modified mean-field $\bar{B}$ function, where the ratio of the screening distance to the particle radius is chosen to provide the best fit at a given value of the surface coverage. An example of this is shown in Figure 3, which contains $\bar{B}$ vs. particle size data for a surface coverage of 1.0%. The analytic $\bar{B}$ function (solid curve) is plotted for a screening distance equal to 29 times the particle radius. This curve provides the best fit near $R = \bar{R}$, where most of the particles in the distribution lie. Chakraverty's function is indicated by the dashed curve, and clearly does not fit the data well. A similar analysis has been performed at other values of the surface coverage by Marsh (7), where the effect of the modified B functions on the asymptotic size distribution are discussed in detail.

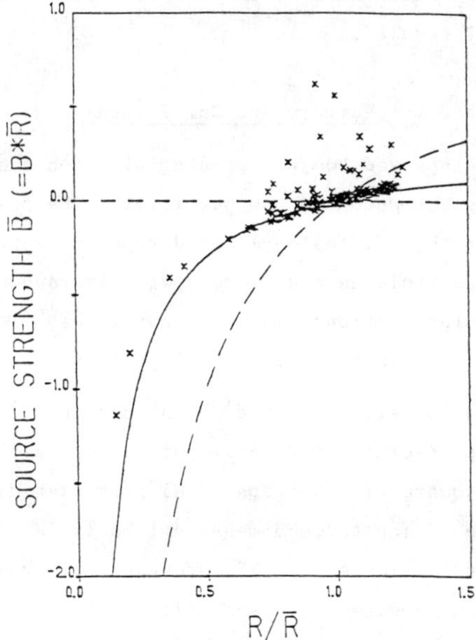

Figure 3 Particle source strengths obtained from two-dimensional simulations at 1.0% surface coverage. The dashed curve shows Chakraverty's mean-field source strength function, whereas the solid curve represents a modified mean-field function using a screening distance equal to 29 times the particle radius.

More extensive computer simulations have been carried out by Voorhees (4) and Voorhees and Glicksman (9), using the three-dimensional diffusion-field model. The main conclusions from these studies may be summarized as follows:

1. Steady-state particle-size distributions have been obtained for a range of volume fractions.
2. The relatively narrow LSW distribution was obtained from the limiting case of zero volume fraction.
3. As the volume fraction of the dispersed phase increases, the distribution becomes broader and somewhat flatter.
4. Over the range of volume fractions examined, the cube of the average radius of the distribution was found to grow linearly with time, once a steady-state distribution was attained.

5. The scaled rate constant for this process approached the LSW value of 4/9 at zero volume fraction, and increased monotonically with increasing volume fraction.
6. These rate constants were compared with those obtained by other continuum ripening theories (9), but unfortunately there is insufficient experimental data at present to differentiate among these theoretical results.

Summary and Conclusions

The aging of a dispersed two-phase material often occurs by the diffusion of heat or mass among the second-phase particles. The present work describes simulation models for this diffusion-limited ripening which determine the growth rate of each particle in a uniform spatially random dispersion directly from the diffusional interactions among the particles. The following is a summary of the multiparticle diffusion model presented in this paper:

1. A finite basis cell is formed which contains N' randomly dispersed particles. For 2-dimensional diffusion, this basis cell is a square with hemispherical particles distributed over the surface. The three-dimensional basis is a cube with spherical particles scattered throughout its volume. The basis cell is then repeated by translation in two or three dimensions as a regular Bravais lattice, to form an infinite, uniform dispersion of particles.
2. Each ripening particle is modeled as an equivalent point source/sink of mass or heat. The resulting diffusion field is equivalent to a dimensionless potential field. Ewald's method is used to derive mathematically convergent forms of both the two- and three-dimensional multiparticle fields.
3. The perimeter- or surface-averaged Gibbs-Thomson boundary condition is applied to each of the N' particles in the basis. This results in N' equations relating the source strengths to the particle sizes and locations in the basis. The overall conservation of mass or heat is used to create one more equation relating the source strengths.
4. The $(N' + 1)$ equations form a linear system which can be solved numerically to determine all of the particle source strengths at any point in time.
5. The growth rate of each particle is mathematically related to its source strength by considering the flux of mass or heat at the particle perimeter or surface. The resulting growth rates are

then used in a time-stepping scheme to simulate the collective ripening of all particles in the system.

The following conclusions can be drawn from the ripening models presented in this paper:

- The multiparticle diffusion field can be recast into a mathematically convergent form through the use of lattice potential techniques.
- Particle growth rates determined by this model arise directly from the diffusional interactions between all particles in a uniform dispersion. Volume-fraction effects are implicitly accounted for through the size and configuration of the basis used in the model.
- Two-dimensional diffusion simulations indicate that the asymptotic particle-size distribution will generally be broader than that predicted by Chakraverty, who used a fixed "screening distance" assumption to determine the individual particle growth rates.
- Extensive computer simulations in three dimensions verify that the LSW distribution and rate constant are obtained in the limiting case of zero volume fraction of the dispersed second phase. Volume-fraction effects are found which indicate broader asymptotic size distributions and higher rate constants result at higher volume fractions.

ACKNOWLEDGEMENT

The authors gratefully acknowledge support of this work by the National Science Foundation under contract DMR83-08052.

References

1. P.P. Ewald, "Die Berechnung optischer und elektrostatischer Gitterpotentiale," Ann. Physik, 64(4) (1921) 253-287.

2. I.M. Lifshitz and V.V. Slyozov, "The Kinetics of Precipitation From Super-saturated Solid Solutions," J. Chem. Phys. Solids, 19 (1961) 35-50.

3. C. Wagner, "Theorie der Alterung von Niederschlagen durch Umlosen, Z. Elektrochem. 65(718) (1961) 581-591.

4. P.W. Voorhees, "Ostwald Ripening in Two-Phase Mixtures" (Ph.D. Thesis, Rensselaer Polytechnic Institute, 1982).

5. J.J. Weins and J.W. Cahn, "The Effect of Size and Distribution of Second Phase Particles and Voids on Sintering", in Sintering and Related Phenomena, ed. G.C. Kuczynski (London: Plenum Press, 1973) 151-63.

6. B.K. Chakraverty, "Grain Size Distribution in Thin Films - 1. Conservative Systems", J. Phys. Chem. Solids, 28 (1967) 2401-2412.

7. S.P. Marsh, "Multiparticle Effects in the Sintering of Supported Metal Catalysts" (M.S. Thesis, Rensselaer Polytechnic Institute, 1984).

8. P.W. Voorhees and M.E. Glicksman, "Solution to the Multi-Particle Diffusion Problem with Applications to Ostwald Ripening - I. Theory", Acta Met., 32(11) (1984) 2001-11.

9. P.W. Voorhees and M.E. Glicksman, "Solution to the Multi-Particle Diffusion Problem with Applications to Ostwald Ripening - II. Computer Simulations", Acta. Met., 32(11) (1984) 2013-2030.

STRESS-INDUCED TRANSFORMATIONS TO COHERENT PRECIPITATES IN NICKEL-BASED ALLOYS

A. F. Jankowski*, E. M. Wingo*, and T. Tsakalakos**

*Rockwell International, North American Space Operations, Rocky Flats Plant, P.O. Box 464, Golden, CO 80402-0464

**Rutgers University, Department of Mech. and Matls. Science, P.O. Box 909, Piscataway, NJ 08854

ABSTRACT

The morphology of precipitates in stress annealed nickel based superalloy crystals has been well documented. Tensile and compressive stress anneals were observed to promote the directional coarsening of the γ' [$Ni_3(Al,Ti)$] precipitates in a γ (nickel solid solution) matrix [1]. Stress induced morphological changes can be completely precluded only if the elastic constants of the matrix and precipitate are equal. It is suggested that the major source of residual strain energy in fully annealed cast nickel-based superalloys is the coherency strain energy associated with the elastic accommodation of misfitting coherent γ' precipitates. In the present investigation, the coherency strain energy Y and the interaction energy δY will be used in a time dependent Fourier representation of the composition profile, assumed periodic cuboidal initially, to describe the experimentally observed change in the γ' morphology under uniaxial stress.

INTRODUCTION

The morphology and orientation of precipitates in a matrix are frequently dominated by strain energy effects. Khachaturyan [2], in calculating the energy of a crystalline precipitate of arbitrary elastic anisotropy in a system consisting of a matrix and precipitate with identical elastic constants, showed that a minimum of the strain energy is consistent with plate-shaped precipitates. Mayo and Tsakalakos [3], utilizing this approach, predicted additional shapes associated with minimum strain energy for precipitates of hexagonal crystal symmetry. Barnett and Asaro [4] furthered Eshelby's method [5] for calculating the strain energy associated with an isotropic precipitate, having an elliposoidal shape, to anisotropic systems. Schneck, Rokhlin, and Dariel [6] have derived a criterion to predict the shape and orientation of a cubic coherent precipitate in a cubic matrix, following a dilational transformation. Lee, Barnett, and Aaronson [7] had previously summarized that only plate-shaped or spherical precipitates were associated with a minimum strain energy for cubic precipitates in a cubic matrix.

In precipitation reactions where coherency is maintained, the elastic strain energy is anisotropic and shape dependent. The elastic energy resulting from the coherency strain plays an important role in the thermodynamics and kinetics of the spinodally decomposed structure as well [8]. Theoretical investigations undertaken by Tsakalakos [9-11] link the morphological changes noted by Tien et al. [1,12], for stress annealed nickel-based superalloy crystals, to the importance of the coherency strain energy. The interaction energy δY, dependent upon the misfit strain and elastic moduli of precipitate and nickel-matrix, has a large contribution to the energy since it may equal as much as 50% of the coherency strain energy Y [9]. The total energy E of the system may be expressed as follows [9]:

$$E = \int [Y(h)+\delta Y(h)]\cdot[\theta(\underline{k})]^2\cdot d\underline{k} \tag{1}$$

where $\theta(\underline{r})$ is a form function which is one inside the precipitate and zero in the matrix, and $\underline{k}$ is the wave vector. $[Y[001]+\delta Y[001]]$ is a minimum for a uniaxial tensile stress [9,11], therefore the total energy E reaches an absolute minimum where $[\theta(\underline{k})]^2$ is a delta-function along [001]. If $\theta(\underline{k})$ is a delta-function in reciprocal space, then the shape of the precipitate defined by $\theta(\underline{r})$ is a delta-function on the plane perpendicular to the [001] direction. Experimental verification from Tien and Copley [1] does indeed show, as seen in Figures 1 and 2, that tensile and compressive stress annealing of Udimet-700 single crystals (14.6-Cr 15.2-Co 4.4-Ti 4.1-Al 4.3-Mo 0.053-C balance-Ni) promoted the directional coarsening of the γ' precipitates in the γ matrix and culminated in precipitate plates with broad faces aligned perpendicular to the stress axis and precipitate parallelepipeds with long axes parallel to the stress axis, respectively. Webster, et. al. [13,14] had observed the effect of applied stress on the periodic morphology of γ'particles in their investigations of polycrystalline Udimet-700.

THEORY

The composition profile, initially assumed to be periodic cuboidal, is shown in Figure 3. To preclude the possibility of γ' shape change due to creep deformation, the stress annealings were carried out at 954°C (1750°C), approximately 80% of the absolute solidus temperature, for 100 hours [1]. The morphology of γ' after annealing without applied stress [1], with the exception of being coarser, remained similar to the initial periodic morphology of cube-aligned cuboids. A Fourier representation of the composition profile will utilize an arithmetic superposition of one-dimensional modulations along the <100> directions [15]. Assuming that the modulations are square composition waves, of amplitude c_0 initially, the composition at a point defined by the Cartesian coordinates (x,y,z) referred to axes along <100> may be represented as

$$\begin{aligned} c(x,y,z,t) = & \left[(a/d)\cdot c_{xo} + \sum_{n_x=1}^{\infty} c_x(k_{nx},t)\cdot\cos(k_{nx}\cdot x)\right] \\ & + \left[(a/d)\cdot c_{yo} + \sum_{n_y=1}^{\infty} c_y(k_{ny},t)\cdot\cos(k_{ny}\cdot y)\right] \\ & + \left[(a/d)\cdot c_{zo} + \sum_{n_z=1}^{\infty} c_z(k_{nz},t)\cdot\cos(k_{nz}\cdot z)\right] \end{aligned} \tag{2}$$

where the component composition wave amplitudes [16] are incorporated into the series representation as

$$c_x(k_{nx},t) = (2/d)\cdot c_{xo}\cdot\frac{\sin(k_{nx}\cdot a)}{k_{nx}}\cdot\exp^{R(k_{nx})\cdot t} \tag{3}$$

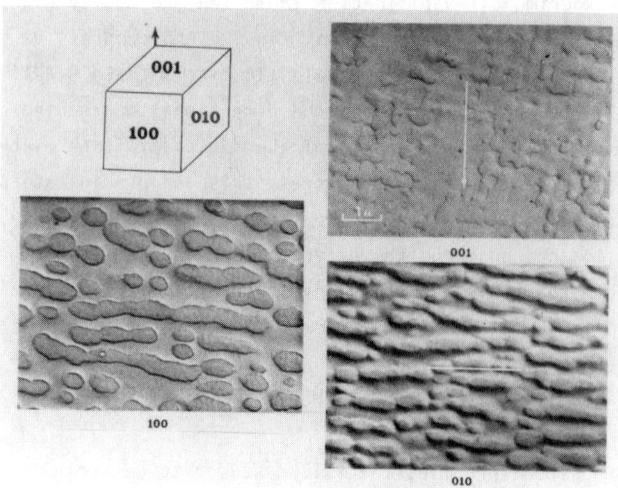

Figure 1. Coarsening of the γ' precipitates in Udimet-700 single crystals for tensile loading [Photo used by permission of J. K. Tien, Met. Trans., 2 (1971) 543-553].

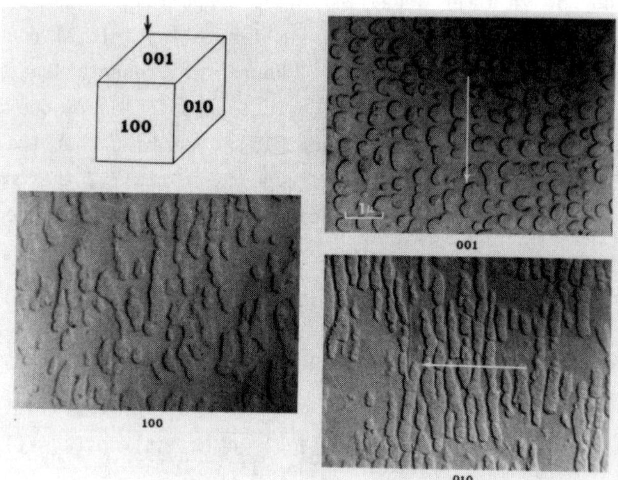

Figure 2. Coarsening of the γ' precipitates in Udimet-700 single crystals for compressive loading [Photo used by permission of J. K. Tien, Met. Trans., 2 (1971) 543-553].

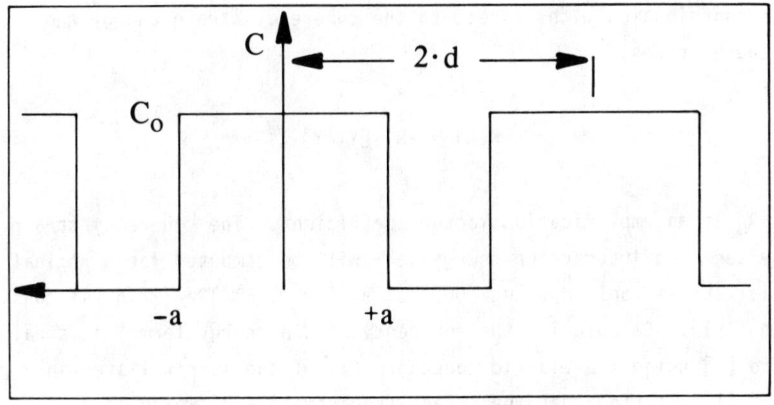

Figure 3. Directional view of the composition wave profile, where the wavelength 2•d = 0.3 μm and 2•a = 0.2 μm [1].

$$c_y(k_{ny},t) = (2/d)\cdot c_{yo}\cdot\frac{\sin(k_{ny}\cdot a)}{k_{ny}}\cdot \exp^{R(k_{ny})\cdot t} \tag{4}$$

$$c_z(k_{nz},t) = (2/d)\cdot c_{zo}\cdot\frac{\sin(k_{nz}\cdot a)}{k_{nz}}\cdot \exp^{R(k_{nz})\cdot t} \tag{5}$$

with wave-numbers $k_{nx} = \pi\cdot n_x/d$; $k_{ny} = \pi\cdot n_y/d$; and $k_{nz} = \pi\cdot nz/d$. The dependency of the composition wave amplitude on the time under uniaxial stress can be directly formulated using a modification of Cahn's linear theory of spinodal decomposition [8]. The amplification factor R(k) of the composition wave amplitude may be expressed as follows:

$$R(k) = -k^2\cdot[\overline{D}+2Kk^2] \tag{6}$$

$$\overline{D} = M\cdot(f''+Y+\delta Y) \tag{7}$$

where $\overline{D}$ is the diffusivity, M is the atomic mobility which is isotropic in cubic crystals, $f'' = \delta^2 f/\delta c^2$, f is the Helmholtz free energy per unit volume, and K is the gradient energy coefficient. Simplifying this expression assuming a negligible gradient energy coefficient and free energy change contribution with respect to the coherency strain energy and interaction energy terms,

$$R(k) = -R_o \cdot [Y+\delta Y] \cdot k^2 \tag{8}$$

where R_o is an amplification factor coefficient. The coherency strain energy term and interaction energy term will be computed for a nominal uniaxial stress applied along [001] of a σ^A = 0.157 TPa (22.5 ksi constant loading) [1]. Calculating the coherency strain energy term Y is straightforward [9] using the elastic constants C_{ij} of the matrix listed in Table I [1]. Noting that the stress-free strain $\varepsilon_o{}^d$ (which is a positive misfit) equals 0.0002 [1],

$$Y\langle 100\rangle = 2 \cdot (\varepsilon_o^d)^2 \cdot [C_{11}+C_{12}-2 \cdot (C_{12}{}^2/C_{11})] = 17.7 \text{ kPa.} \tag{9}$$

The interaction energy δY [9] for both compressive and tensile loads are easily computed and are listed in Table II, where

$$\delta Y[001] = \sigma^A \cdot \varepsilon_o^d \cdot \frac{(\bar{C}_{11}+\bar{C}_{12}) \cdot \delta C_{11} - 2 \cdot \bar{C}_{12} \cdot \delta C_{12}}{\bar{C}_{11} \cdot (\bar{C}_{11}-\bar{C}_{12})} - \frac{\delta C_{11}+2 \cdot \delta C_{12}}{\bar{C}_{11}+2 \cdot \bar{C}_{12}} \tag{10}$$

and

$$\delta Y[001] = \delta Y[010] = \sigma^A \cdot \varepsilon_o^d \cdot \frac{\bar{C}_{11} \cdot \delta C_{12} - \bar{C}_{12} \cdot \delta C_{12}}{\bar{C}_{11} \cdot (\bar{C}_{11}-\bar{C}_{12})} - \frac{\delta C_{11}+2 \cdot \delta C_{12}}{\bar{C}_{11}+2 \cdot \bar{C}_{12}} \tag{11}$$

Incorporating the interaction energies of Table II and the energy coefficients of Table III, noting

$$\alpha = Y\langle 100\rangle + \delta Y[001] \tag{12}$$

and

$$\beta = Y\langle 100\rangle + \delta Y[100] \tag{13}$$

the amplification factors $R(k_n)$ of Equations (3) to (5) may be rewritten

$$R(k_{nx}) = -(k_{nx})^2 \cdot R_o \cdot \beta \tag{14}$$

TABLE I. ELASTIC CONSTANTS [1]

C_{ij}	C_{11} (TPa)	C_{12} (TPa)	C_{44} (TPa)
Matrix	.2508	.1500	.1235
Precipitate	.2148	.1277	.0997
δ(difference)	-.0360	-.0233	-.0238

TABLE II. INTERACTION ENERGIES [9]

State of Stress	δY[001] (kPa)	δY[100]=δY[010] (kPa)
Tension σ_T^A	-4.836	4.420
Compression σ_C^A	4.836	-4.420

TABLE III. ENERGY COEFFICIENTS

State of Stress	α (kPa)	β (kPa)	α/β
Tension σ_T^A	12.86	22.12	.58
Compression σ_C^A	22.54	13.28	1.70

$$R(k_{ny}) = -(k_{ny})^2 \cdot R_o \cdot \beta \qquad (15)$$

$$R(k_{nz}) = -(k_{nz})^2 \cdot R_o \cdot \alpha \qquad (16)$$

A time increment δt must be determined to clearly manifest the morphological changes of the stress-induced transformation.

$$t = m \cdot \delta t \qquad m = 1,2,3... \qquad (17)$$

Assuming an $R(k_{nx}) \cdot t$ value of -0.1 would assuredly evidence the experimentally observed changes over several time increments $m \cdot \delta t$ for a tensile loading along [001]. Utilizing the α and β values of Table III (noting that δY[100] is approximately 25% of Y<100>) and Equations (15) and (18), the expression for the time increment is derived equating $R(k_{nx}) \cdot t$ for:
$n_x=1$; $m=1$

$$\delta t = (R_o \cdot 10^5)^{-1} . \tag{18}$$

The stress aging induced directional coarsening of the γ'-precipitates may now be computer simulated using Equations (2) and (5) and the following expressions of $R(k_n) \cdot t$

$$R(k_{nx}) \cdot t = -(k_{nx})^2 \cdot \beta \cdot m \cdot 10^{-5} \tag{19}$$

$$R(k_{ny}) \cdot t = -(k_{ny})^2 \cdot \beta \cdot m \cdot 10^{-5} \tag{20}$$

$$R(k_{nz}) \cdot t = -(k_{nz})^2 \cdot \alpha \cdot m \cdot 10^{-5} \tag{21}$$

DISCUSSION

The differentiation of the reaction path in the formation of a periodic array of equilibrium concentration particles, between nucleation-and-growth [17,18] and spinodal decomposition [19,20], has continued under several recent investigations [21,22]. In the present study, the induced directional coarsening of the γ' [$Ni_3(Al,Ti)$] precipitates (after stress aging) has been simulated utilizing a composition wave profile based on the theory of spinodal decomposition [8,23]. The increments of time selected clearly show the initial, intermediate, and final structures (as seen in Figures 4 to 8) during the stress-induced directional coarsening for both uniaxial (compressive and tensile) loadings. A value for the time increment δt may be calculated, if the diffusivity is known, by combining Equations (6), (7), (8), and (18), and adhering to the underlying assumptions in their derivation. Approximating an interdiffusion coefficient of $\bar{D} = 1 \times 10^{-19}$ m^2/s [22,24], the time increment δt used is roughly 30 minutes. Therefore, $t = 100 \cdot \delta t$ corresponds to 50 hours, which is on the same scale as experimentally reported [1].

SUMMARY

A major source of residual strain energy is the coherency strain energy associated with the elastic accommodation of misfitting coherent γ'-precipitates in a γ-matrix. The coherency strain energy Y and interaction energy δY have been incorporated into a time dependent Fourier representation of the composition to predict the experimentally observed changes in the γ' morphology of a nickel-based superalloy.

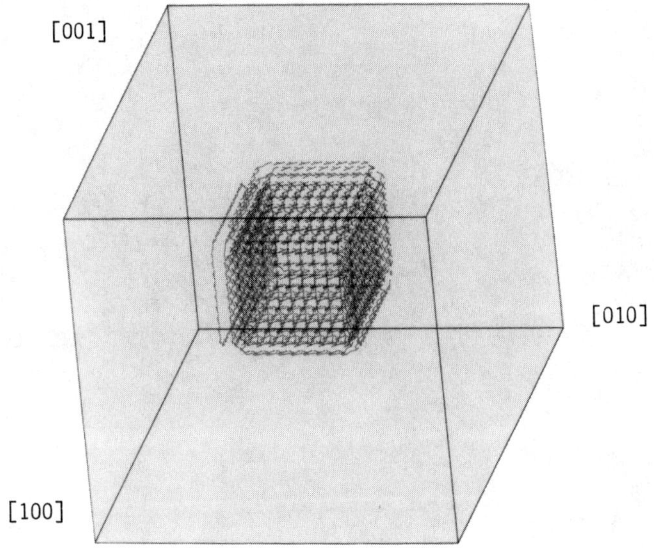

Figure 4. A three-dimensional computer representation of the initial cuboidal structure, t = 0. Composition contour lines c(x,y,z,t)=90% are shown.

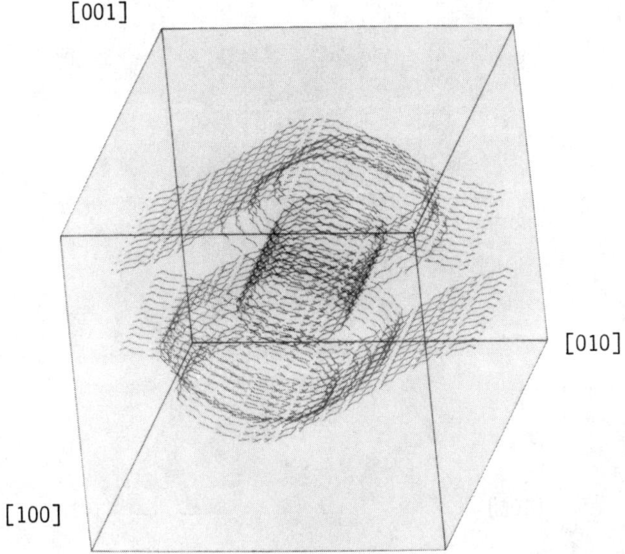

Figure 5. Coarsening of the structure under [001] tensile loading at time t = 10 δt. Composition contour lines c(x,y,z,t)=60%,90% are shown.

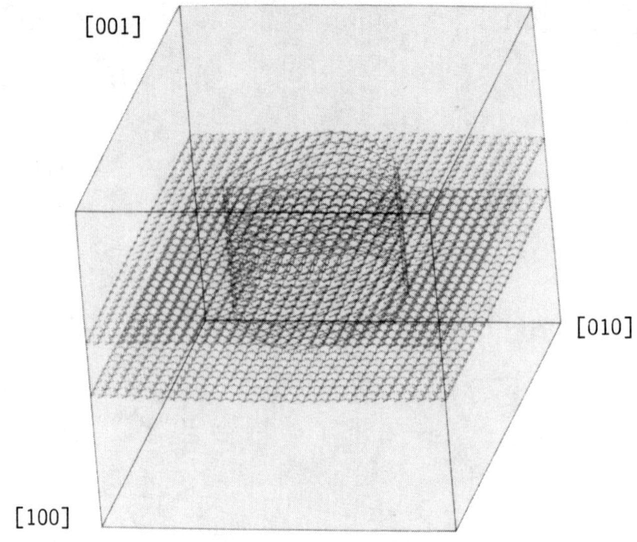

Figure 6. Coarsening of the structure under [001] tensile loading at time $t = 100\ \delta t$. Composition contour lines $c(x,y,z,t)$=90% are shown.

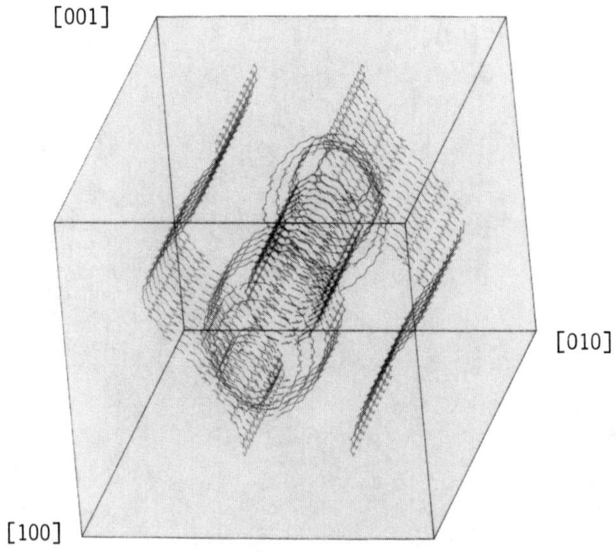

Figure 7. Coarsening of the structure under [001] compressive loading at time $t = 10\ \delta t$. Composition contour lines $c(x,y,z,t)$=60%,90%.

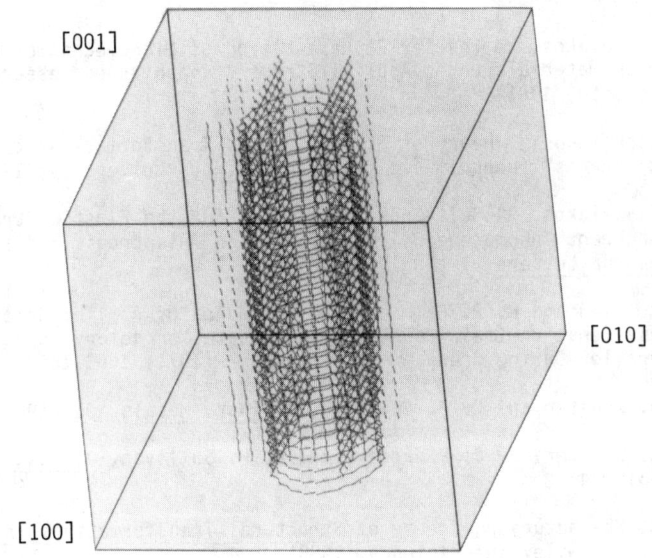

Figure 8. Coarsening of the structure under [001] compressive loading at time t = 100 δt. Composition contour lines c(x,y,z,t)=90% are shown.

REFERENCES

1. J. K. Tien and S. M. Copley, "The Effect of Uniaxial Stress on the Periodic Morphology of Coherent Gamma Prime Precipitates in Nickel-Base Superalloy Crystals," Met. Trans., 2 (1971) 215-219; "The Effect of Orientation and Sense of Applied Uniaxial Stress on the Morphology of Coherent Gamma Prime Precipitates in Stress Annealed Nickel-Base Superalloy Crystals," Met. Trans., 2 (1971) 543-553.

2. A. G. Khachaturyan, Soviet Phys. Solid State, 8 (1967) 2163.

3. W. E. Mayo and T. Tsakalakos, "The Influence of Elastic Strain Energy on the Formation of Coherent Hexegonal Phases," Met. Trans., 11A (1980) 1637-1644.

4. R. J. Asaro and D. M. Barnett, J. Mech. Phys. Solids, 23 (1975) 77.

5. J. D. Eshelby, Prog. Solid Mech., 2 (1961) 89.

6. R. Schneck, S. I. Rokhlin, and M. P. Dariel, "Criterion for Predicting the Morphology of Crystalline Cubic Precipitates in a Cubic Material," Met. Trans., 16A (1985) 197-202.

7. J. K. Lee, D. M. Barnett, and H. I. Aaronson, "The Elastic Strain Energy of Coherent Ellipsoidal Precipitates in Anisotropic Crystalline Solids," Met. Trans., 8A (1977) 963-970.

8. J. W. Cahn, "On Spinodal Decomposition in Cubic Crystals," Acta Met., 10 (1962) 179-183; "Coherent Fluctuations and Nucleation in Isotropic Solids," Acta Met., 10 (1962) 907-913.

9. T. Tsakalakos, "A Fourier Representation of Micromechanics of Inhomogeneous Materials Under Applied Stress" (unpublished research, Rutgers University, 1981).

10. T. Tsakalakos, "Theory of Stress-Induced Transformations to Coherent Precipitates" (unpublished research, Rutgers University, 1971).

11. T. Tsakalakos, "A Self-Consistent Model for the Elastic Strain Energy of Coherent Inhomogeneous Precipitates in Anisotropic Media," Res. Mechanica Letters, 1 (1981) 409-415.

12. J. K. Tien and R. P. Gamble, "The Influence of Applied Stress and Stress Sense on Grain Boundary Precipitate Morphology in a Nickel-Base Superalloy During Creep," Met. Trans., 2 (1971) 1663-1667.

13. G. A. Webster and C. P. Sullivan, J. Inst. Metals, 95 (1967) 138.

14. G. A. Webster, B. J. Piearcey, and C. P. Sullivan, J. Inst. Metals, 96 (1968) 274.

15. A. G. Khachaturyan, Theory of Structural Transformations in Solids (New York: J. Wiley Interscience, 1983), 324.

16. J. W. Cahn, "On Spinodal Decomposition," Acta Met., 9 (1961) 795-801.

17. R. O. Williams, "Aging of Nickel Base Aluminum Alloys," Trans. Amer. Inst. Mining Engrs., 215 (1959) 1026-1032.

18. A. J. Ardell and R. B. Nicholson, "On the Modulated Structure of Aged Ni-Al Alloys," Acta Met., 14 (1966) 1295-1309.

19. C. L. Corey, B. Z. Rosenblum, and G. M. Greene, "The Ordering Transition in Ni_3Al Alloys," Acta Met., 21 (1973) 837-844.

20. W. O. Gentry and M. E. Fine, "Precipitation in Ni - 11.1 at.% Al and Ni - 13.8 at.% Al Alloys," Acta Met., 20 (1972) 181-190.

21. P. Haasen, "The Early Stages of the Decomposition of Alloys," Met. Trans., 16A (1985) 1173-1184.

22. S. A. Hill and B. Ralph, "Continuous Phase Separation in a Nickel-Aluminum Alloy," Acta Met., 30 (1982) 2219-2225.

23. J. W. Cahn, "Phase Separation by Spinodal Decomposition in Isotopic Systems," J. Chem. Phys., 42 (1965) 93-99.

24. A. M. Irisarri, J. J. Urcola, and M. Fuentes, "Kinetics of growth of γ'-precipitates in Ni - 6.75 Al alloy," Matls. Sci. Tech., 1 (1985) 516-519.

COMPUTER SIMULATION OF DISLOCATION PRODUCTION AND RECOVERY FOR IRRADIATED METALS

Charles H. Henager, Jr. and E. P. Simonen

Pacific Northwest Laboratory
Richland, WA 99352

Abstract

The effect of the thermal stability of interstitial dislocation loops on the hardening-softening transition temperature of irradiated metals was explored with a simple computer model. Destabilizing the interstitial loop population by simulating spontaneous thermally-activated unfaulting reduced the hardening-softening transition temperature significantly. Simulated loop unfaulting shifted the transition temperature down 50 K for both annealed and cold-worked microstructures. Interstitial loop growth, loop coalescence with network dislocations, and network recovery were simulated using rate equations. Simple expressions for dipole recovery, dislocation intersections, and loop population behavior were developed. The model correctly predicts typical trends observed in irradiated metals. These trends include saturation of the dislocation density at high irradiation doses and a decrease in loop density with increasing irradiation temperature. However, the model underestimates the total loop densities and, therefore, the magnitude of the predicted strengths and softening temperatures are less than observed.

Introduction

Interstitial loop growth causing radiation hardening dominates the evolution of microstructure during irradiation of metals for temperatures below about one-half the melting temperature, 0.5Tm. Recovery processes causing softening dominate above 0.5Tm. The transition temperature from hardening to softening is an important parameter for reactor structural materials. High temperature strength and creep resistance are determined by the resistance of the material to recovery processes. Hardening-softening transitions have been determined for austenitic and ferritic steels in neutron environments (1-3). The transition from hardening to softening occurs at about 773 K in austenitic steels (1) and about 723 K in ferritic steels (2,3). The hardening mechanism in the case of austenitic steels was related to a combination of dislocation loops and network dislocations (1), while the ferritic steels exhibit additional hardening due to precipitation and solution hardening (2). Dislocation recovery was stated as the cause of the loss of strength at high temperatures in each case.

Uncertainties in loop production and recovery mechanisms make it difficult to calculate the transition from radiation hardening to softening. Loop nucleation is difficult to model, but loop recovery can be estimated by calculating the temperature at which thermal vacancy capture destabilizes the loop population (4). However, predicted softening transition temperatures for irradiated ferritic steels are as much as 50 to 75 C higher than observed softening temperatures (4). This suggests that loop stability is an issue and is being overestimated by thermal vacancy capture only.

Interstitial dislocation loops grow by interstitial capture, shrink by vacancy capture, and unfault due to thermal fluctuations, collisions with network dislocations and other loops. Sessile faulted dislocation loops are very effective dislocation barriers and are more resistant to recovery processes than network dislocations, which can climb and glide under local stresses to recover in a dipole geometry. Loop unfaulting transforms the sessile loop into a glissile loop and greatly reduces its effectiveness as an obstacle. The unfaulted loop is also more susceptible to recovery since it can accommodate local stresses by glide and will coalesce with the network more easily. Faulted loop stability and the hardening-softening transition are related in a complex way that depends on the details of the loop-network recovery process.

A simple model has been formulated to simulate the effects of loop stability on the hardening-softening transition of an irradiated metal. Loop growth and evolution of the network dislocation microstructure due to loop-network coalescence and network recovery are described by the model. Hardening due to faulted loops and network dislocations was calculated as a function of irradiation dose and temperature. The effect of destabilizing the faulted loop, allowing it to thermally unfault and join the network, was to shift the hardening-softening transition temperature down about 50 K to lower temperatures. Predicted transition temperatures were less than observed due to underestimating the total loop density in the model. Network dislocation densities predicted by the model at a constant temperature saturated as observed in irradiated metals, although the predicted saturation densities were less than typical observations. Loop densities were predicted to decrease with increasing irradiation temperature.

Model Formulation

Reaction rate theory was used to calculate interstitial and vacancy concentrations, loop growth rates, and network dislocation climb velocities (5). The following four equations were used for these calculations:

$$\frac{dC_i}{dt} = 0 = K - \alpha C_i C_v - D_i C_i [Z_i \rho_d(t) + 4\pi r_{c,i} \rho_\ell(t) + 4\pi r_c \rho_c] \quad (1)$$

$$\frac{dC_v}{dt} = 0 = K - \alpha C_i C_v - D_v C_v [Z_v \rho_d(t) + 4\pi r_{c,v} \rho_\ell(t) + 4\pi r_c \rho_c]$$
$$+ D_v [C^e_{v,d} Z_v \rho_d(t) + C^e_{v,\ell} 4\pi r_{c,v} \rho_\ell(t) + C^e_{v,c} 4\pi r_c \rho_c] \quad (2)$$

$$\frac{dr_\ell}{dt} = \frac{2}{r_\ell b} [r_{c,i} D_i C_i - r_{c,v} D_v (C_v - C^e_{v,\ell})] \quad (3)$$

$$V = \mathrm{ABS} \left\{ \frac{1}{b} [Z_i D_i C_i - Z_v D_v (C_v - C^e_{v,d})] \right\} \quad (4)$$

The single subscripts refer to interstitial (i), vacancy (v), dislocation (d), loop (l), and cavity (c). $r_{c,v}$ and $r_{c,i}$ refer to capture radii of interstitial loops for vacancies and interstitials. C^e_v refers to the equilibrium concentration of vacancies. The vacancy concentration in equilibrium with dislocations ($C^e_{v,d}$), loops ($C^e_{v,\ell}$), or cavities ($C^e_{v,c}$) is also computed. The defect production rate (K), biases (Z), and diffusivities (D) are defined in Table I.

Table I. Rate Theory Parameters

Parameter	Symbol	Value
Shear Modulus	G	9.5×10^{10} Pa
Burgers Vector	b	2.5×10^{-10} m
Recombination Constant	α	$1.2 \times 10^{21}(D_i + D_v)\ s^{-1}$
Equil. Vacancy Conc.	C^e_v	$\exp(1.5)\exp(-1.7/kT)$
Vacancy Diffusivity	D_v	$2 \times 10^{-5} \exp(-1.3/kT)\ m^2/s$
Interstitial Diffusivity	D_i	$1 \times 10^{-4} \exp(0.3/kT)\ m^2/s$
Dislocation Bias Factors:		
Vacancy	Z_v	$2\pi/\ln(\lambda_d/2b)$, $\lambda_d = (\pi\rho_d)^{-1/2}$
Interstitial	Z_i	$Z_v(1 + \Delta Z_i)$
Stress-Free Interstitial Bias	ΔZ_i	0.10
Defect Production Rate	K	1×10^{-6} dpa/s

The model assumes that dislocation loops nucleate, grow, and coalesce with network dislocations. The rate at which coalscence occurs depends on the loop growth rate and the loop size at which coalescence becomes highly probable. The average loop growth rate was computed in lieu of computing the evolution of the entire loop size distribution. The transit time for a loop to grow from an infinitesimal faulted loop to a size where it becomes part of the Frank network was used as the indication of loop stability. Two geometrical arguments, loop-loop and loop-network intersections, define the largest average loop that can survive as a faulted loop. Spontaneous thermally-activated loop unfaulting was also considered as a mechanism for a loop to make a transition to the network. At high temperatures a loop may unfault by nucleation of the appropriate partial dislocation. The resulting glissile loop is considered to rapidly join the network by a glide and coalescence reaction.

A critical loop radius was defined as a radius above which a loop transforms from a loop to the network. For a sparse network, loops will intersect one another, unfault, and coalesce to form a network. The critical radius, r_ℓ^c, for loop-loop intersections satisfies the following space-filling relationship:

$$\frac{4}{3}\pi(r_\ell^c)^3\rho_\ell = 1 \tag{5}$$

For a dense network, the critical radius will be one-half the average network dislocation spacing.

$$r_\ell^c = 0.5(\pi\rho_d)^{-1/2} \tag{6}$$

A critical radius was calculated for both sparse and dense network dislocation densities.

Thermally-activated loop unfaulting has been discussed by Saada (6), but it requires a large activation energy even for metals with large stacking fault energies unless unfaulting is assisted by high local stresses. The driving force for loop unfaulting is removal of the stacking fault and the activation energy is related to the stacking fault energy. As shown in Figure 1, the predicted activation energies for spontaneous loop unfaulting are very large based on the analysis of Saada (6). Nevertheless, there are observations of loop unfaulting for metals with high stacking fault energies that indicate that the activation energy is overestimated by this analysis, implying that most loop unfaulting is probably stress-assisted (7-9). Metals with low stacking fault energies, such as austenitic steels, are observed to contain large, faulted dislocation loops that appear to be resistant to spontaneous unfaulting (10).

Two extreme cases were considered for this paper to explore the role of loop unfaulting on loop stability. One case restricted loop lifetimes by the geometric intersections only, simulating a metal with a low stacking fault energy for which thermally-activated loop unfaulting is unlikely. The other extreme is one for which the critical radius for loop unfaulting is a strong function of temperature and is less than that for the geometrical arguments, simulating a metal with a high stacking fault energy. An arbitrary temperature-dependence of the critical loop radius was defined to

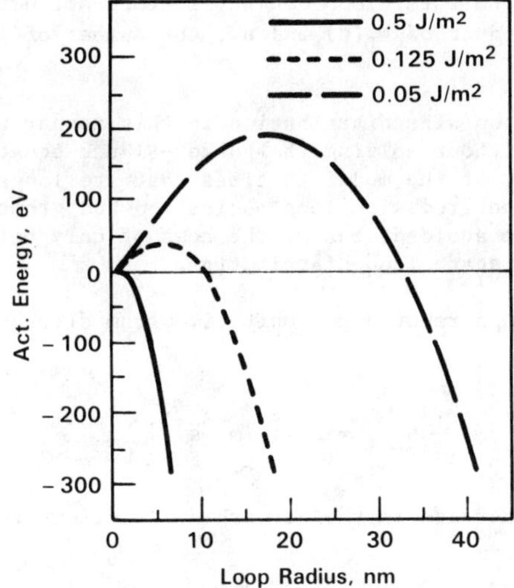

Figure 1 - Activation energy for spontaneous dislocation loop unfaulting as a function of the indicated stacking fault energy. Curves were calculated based on the analysis of Saada (6). The curve for 0.5 J/m^2 represents the activation energy for an "effective" stacking fault energy as discussed in reference 7.

simulate loop unfaulting at sizes below the geometric critical radius. This was necessary due to the lack of an analytical expression for the critical radius with spontaneous loop unfaulting and the inability to predict loop unfaulting kinetics.

Microstructural evolution was predicted by simulating the nucleation and growth of a continuous loop distribution using discrete loop distributions. The time-dependence of the discrete loop distributions was computed based on the growth rate of the average loop in the distribution. An assumed loop distribution was partitioned into a number of smaller distributions each having a lifetime, τ, from nucleation to coalescence with the network. The discrete loop distribution lifetime was determined by computing the time it took the average loop in the distribution to reach the critical loop radius. The total loop distribution is allowed to evolve based on the production and annihilation of the discrete loop distributions. In this manner, a continuous flow of dislocation loops nucleating, growing, unfaulting and coalescing is simulated by the model. The network dislocation density is allowed to change to reflect the contribution due to dislocation loop growth and coalescence, as well as to reflect dislocation annihilation through network recovery.

A simple differential equation describes the time dependence of the total loop distribution:

$$\frac{dN_D}{dt} = \frac{dP_D}{dt} - \frac{N_D}{\tau} \; ; \; N_D = \dot{P}_D \tau (1 - e^{-t/\tau}) \qquad (7)$$

where N_D is the number of discrete loop distributions in transit, $\dot{P}_D$ is the production rate of discrete loop distributions, and τ is the discrete distribution lifetime. The total loop density is then a function of time with $\rho_\ell(t)$ given by the product of $N_D(t)$ and n_ℓ, the number of loops in each discrete distribution.

Simulating the loop size distribution in this manner allows the modeling to be performed without solving the Fokker-Planck equation (11), but restricts the validity of the model to times that are long compared to τ. The difficulties encountered with loop nucleation and growth of the loop size distribution were avoided, but at the cost of only being able to predict the quasi steady state loop distribution.

The number of loops removed per unit time from discrete loop distributions is given by:

$$\frac{1}{\tau} N_D n_\ell = \rho_\ell/\tau \ [m^{-3}s^{-1}] \tag{8}$$

which results in an increase in the network dislocation line length given by:

$$(2\pi r_\ell^c \rho_\ell)/\tau \ [m^{-2}s^{-1}] \tag{9}$$

since the loops have an average size r_ℓ^c after a lifetime τ.

Network recovery competes with the network build-up through climb-assisted dislocation annihiliation (12). Network dislocations are assumed to be arranged in a dipole geometry having an average separation of h. Opposite signed dislocations climb in opposite directions with velocity V which results in an average recovery time, τ_R, given by:

$$\tau_R = \frac{h}{2V} \ ; \quad h = (\pi \rho_d)^{-1/2} \tag{10}$$

The rate of recovery is proportional to the dipole density, $\rho_d/2$, divided by the average recovery time:

$$\frac{\rho_d}{2\tau_R} = \frac{\rho_d V}{h} = \pi^{1/2} \ \rho_d^{\ 3/2} V \ \ [m^{-2}s^{-1}] \tag{11}$$

Network source and loss terms are combined to give:

$$\frac{d\rho_d}{dt} = \frac{1}{\tau} \ \rho_\ell 2\pi r_\ell^c - \pi^{1/2} \rho_d^{\ 3/2} V \tag{12}$$

for an expression for the evolution of the network dislocation density.

Simple hardening laws yield expressions to calculate changes in flow stress based on loop and network densities. The flow stress is given by (1):

$$\sigma = Gb(\rho_\ell \bar{r}_\ell)^{1/2} + \frac{Gb}{4} \rho_d^{1/2} \qquad (13)$$

where the first term is the strength contribution from dislocation loops having a time-averaged radius, $\bar{r}_\ell$, and the second term is the strength contribution from network dislocations.

Model Implementation

A FORTRAN program was written to perform the computations for the model. Initial parameter values are given in Table 2. Computations were performed at 473, 573, 673, 773, and 823 K to span the temperatures from irradiation hardening dominance to softening dominance. At T = 473 K, the average loop lifetime was calculated as a function of ρ_ℓ and ρ_d. Since ρ_ℓ and τ are related by equation 7 and are not independent, it was necessary to select self-consistent values for ρ_ℓ and τ at the start of the computation. This fixed $\rho_\ell(t=0)$ and τ for the given value of $\rho_d(t=0)$. $\rho_d(t=0)$ was either 3×10^{14} m^{-2} or 1×10^{12} m^{-2} to represent cold-worked or annealed material. Typical starting values of ρ_ℓ was 2×10^{20} m^{-3} after iterating to find ρ_ℓ and τ at t=0. $N_D(t=\infty)$ was given by equation 7 and n_ℓ was set equal to $1 \times 10^{22} / N_D(t=\infty)$.

Table 2. Initial Values

Parameter	Symbol	Value
Network Density	ρ_d	3×10^{14} or 1×10^{12} m^{-2}
Loop Density	ρ_ℓ	$N_D(t=\infty) \cdot n_\ell$
Cavity Density	ρ_c	5×10^{21} m^{-3}
Loop Population Creation Rate	$\dot{P}_D$	1×10^{-4} s-1

Time increments were controlled to keep variations in ρ_d less than 10% in a single time step consistent with maintaining the largest possible time step. The ρ_d, τ, ρ_ℓ, N_D, and σ were updated at each time step, while the n_ℓ was held constant at the value computed at t=0. In this manner, $\rho_\ell(t)$ was constrained to be less than or equal to 1×10^{22} m^{-3}.

The same input parameters were assumed at each temperature with the exception of n_ℓ, the number of loops per discrete distribution. In this manner, the model allowed ρ_ℓ to vary as a function of temperature depending on the values of τ calculated. Loop lifetimes and growth rates are a function of temperature through the dependence of interstitial-vacancy recombination on temperature. Recombination decreases as vacancies become more mobile. This temperature effect caused ρ_ℓ to decrease with increasing temperature, simulating a common observation in irradiated metals.

Results and Discussion

Loop stability assumptions have a strong influence on the hardening-softening transition temperature, Figure 2. Loop unfaulting caused by thermal activation at sizes below those for the calculated geometric

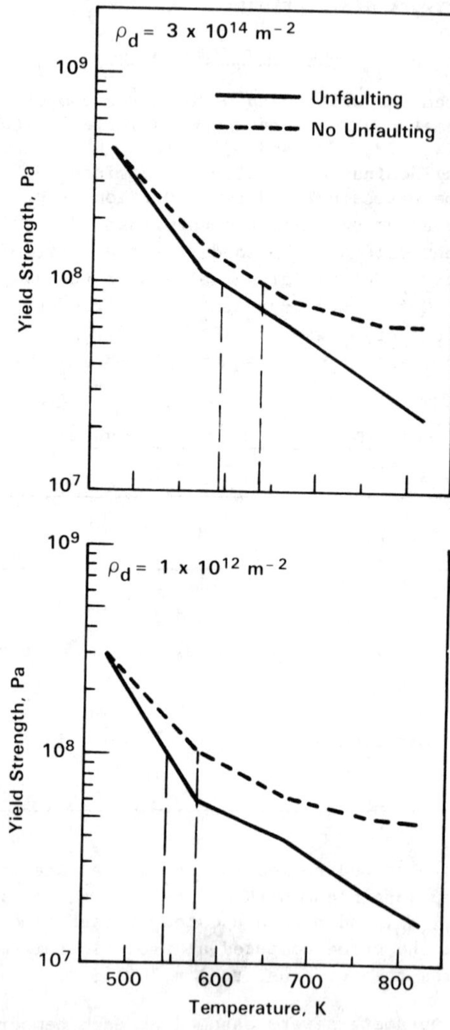

Figure 2 - Calculated yield strengths of an irradiated metal based on loop unfaulting or no loop unfaulting. The curves are for cold-worked, $\rho_d = 3x10^{14} m^{-2}$, and annealed, $\rho_d = 1x10^{12} m^{-2}$, starting assumptions. The temperatures are indicated for which a yield strength of 100 MPa is calculated after 50 dpa.

intersections, equations 5 and 6, was simulated by assuming a reduced critical loop radius. Shifts in the transition temperature to lower temperatures were calculated. The predicted temperature for a flow stress of 100 MPa was shifted down 50 K for the case of loop unfaulting compared to no unfaulting. This downward shift in the transition temperature is because faulted loops recover less easily than do network dislocations.

Loop lifetimes computed by the model, shown in Table 3, illustrate the effects of temperature and simulated loop unfaulting on the lifetime of the average dislocation loop. Short loop lifetimes lead to rapid recovery rates

Table 3. Loop Lifetimes (in seconds)

	$\rho_d = 3 \times 10^{14}\ m^{-2}$		$\rho_d = 1 \times 10^{12}\ m^{-2}$	
T(K)	no unf.	unf.	no unf.	unf.
473	1.6×10^6 s	1.6×10^6	7.7×10^6	7.7×10^6
573	6.4×10^4	6.4×10^4	6.0×10^5	1.0×10^5
673	1.4×10^4	1.4×10^4	1.5×10^5	7.7×10^3
773	9.8×10^3	8.4×10^3	7.6×10^4	1.4×10^3
823	9.7×10^3	4.0×10^3	6.6×10^4	4.8×10^2

and cause a reduction in yield stress due to the reduction of dislocation line length. Loop lifetimes are also sensitive to loop density. Low loop densities lead to short lifetimes and rapid recovery rates.

Predicted hardening-softening transition temperatures of about 600 K, based on achieving a flow stress of 100 MPa, were less than the observed transition temperatures of 700 K to 800 K (1-3). This suggests that recovery was too rapid in the model. Recovery rates were overestimated because predicted loop densities were lower than observed densities. Low loop densities cause insufficient hardening and also increase the growth rate of the loops so that loop lifetimes are short. Uncertainties in selection of P_D may have contributed to underestimating loop densities, but the loop lifetimes, especially at 473 K, were determined by r_ℓ^c as ρ_d increased.

Refinement of the Frank network in metals typically results in dislocation cell formation, which is not accounted for in this model. Loops would suffer more loop-loop interactions and less loop-network interactions in a cell structure. The validity of equation 6 is restricted to a Frank network and would not be valid for a cell structure. Describing cell formation at high dislocation densities makes the problem of microstructural evolution much more difficult.

Detailed model results agree with trends observed in irradiation testing of metals, Figures 3 and 4. Loop density decreases as a function of temperature have been observed during light-ion and neutron irradiations (13,14). Saturation of the dislocation density has been observed for austenitic and ferritic steels after about 10 to 20 dpa (12). The rapid recovery occurring in the model causes saturation to occur at doses lower than typical observations. The predicted saturation density decreases with

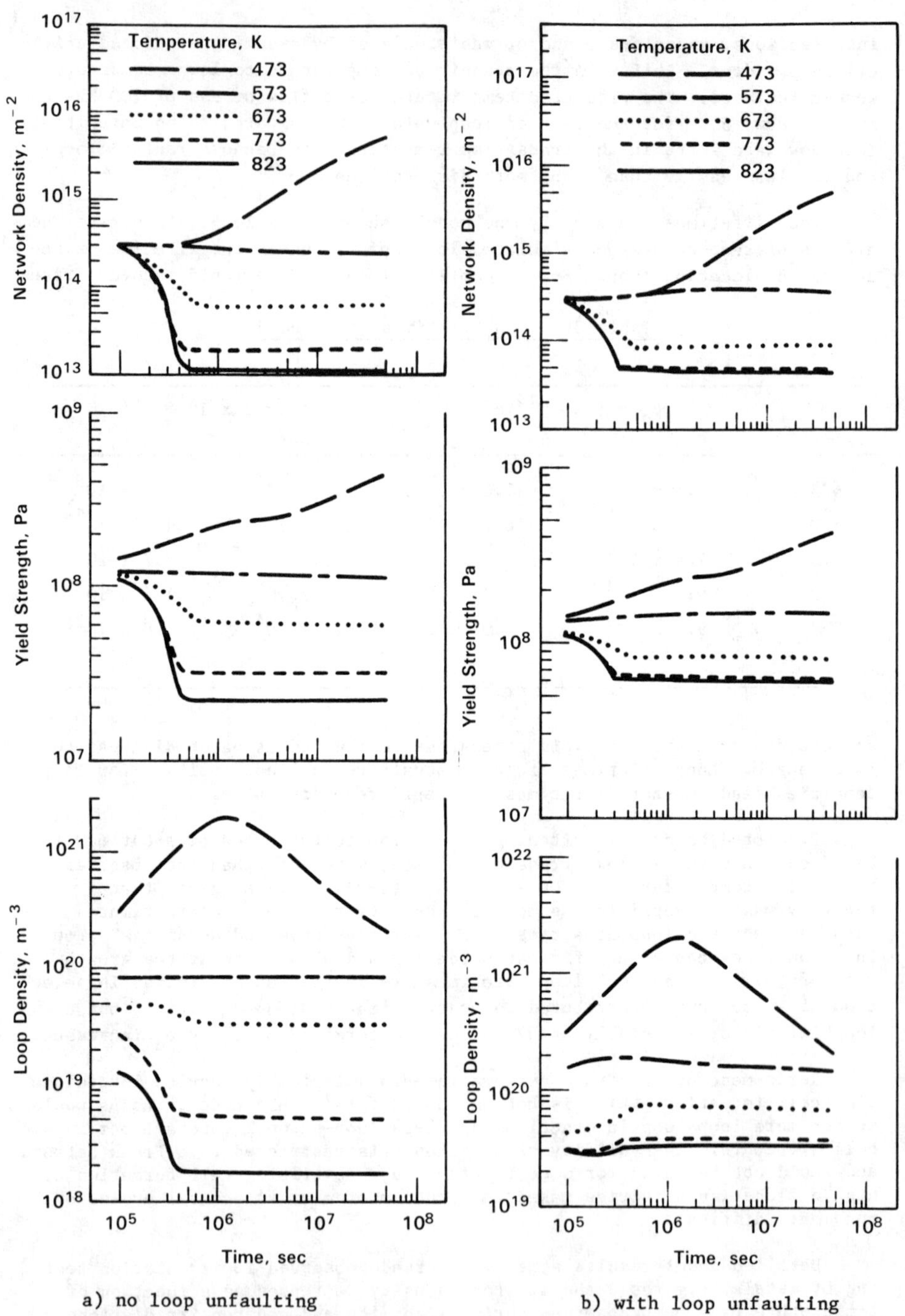

Figure 3 - Model predictions of network dislocation density, yield strength, and dislocation loop density as functions of irradiation temperature and irradiation time for a starting dislocation density of $\rho_d=3x10^{14}m^{-2}$. Total dose is 50 dpa.

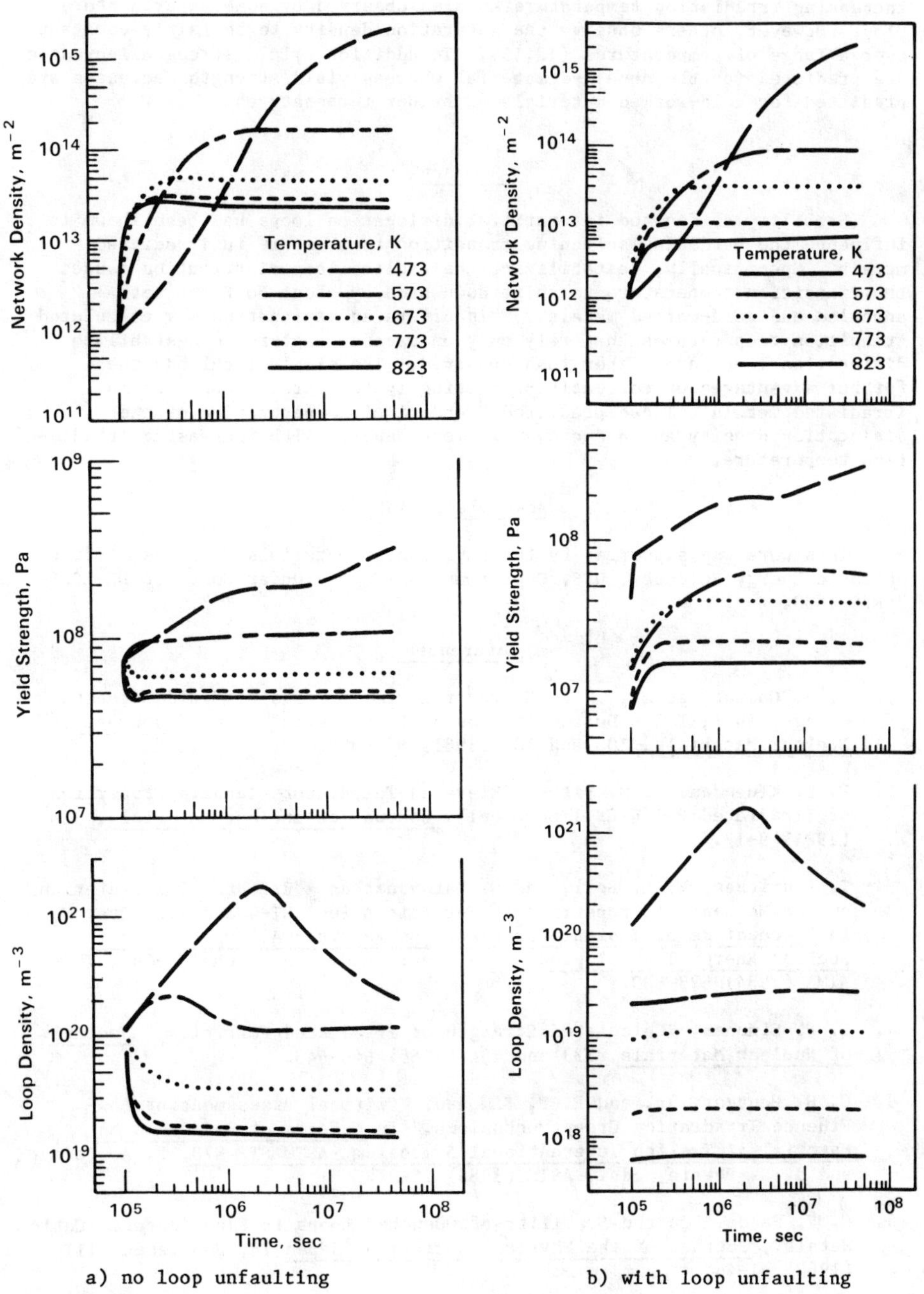

a) no loop unfaulting

b) with loop unfaulting

Figure 4 - Model predictions of network dislocation density, yield strength, and dislocation loop density as functions of irradiation temperature and irradiation time for a starting dislocation density of $\rho_d = 1 \times 10^{12} m^{-2}$. Total dose is 50 dpa.

increasing irradiation temperature as also observed by some experimenters (15). However, others observe the saturation density to be fairly constant over a range of temperatures (12,15). In addition, yield strength increases are predicted for the annealed material whereas yield strength decreases are predicted for cold-worked materials at higher temperatures.

Summary

Stability of faulted interstitial dislocation loops has been shown to influence the hardening-softening transition temperature in irradiated metals. Specifically, destabilizing the faulted loop distribution causes the transition temperature to shift downwards by about 50 K for both annealed and cold-worked metals. This offers an explanation why calculated transition temperatures that rely only on vacancy capture to destabilize dislocation loops are higher than observed. The simple model has the further advantages of correctly predicting typical trends observed in irradiated metals. These predicted trends include saturation of the dislocation density and a decrease in loop density with increasing irradiation temperature.

Acknowledgement

This work was supported by the Division of Materials Sciences, Office of Basic Energy Sciences, U.S. Department of Energy under Contract DE-AC06-76RLO-1830.

References

1. F. A. Garner, et al., "The Microstructural Origins of Yield Strength Changes in AISI 316 During Fission or Fusion Irradiation," Journal of Nuclear Materials, 103 and 104 (1981) 803-808.

2. R. L. Klueh and J. M. Vitek, "Elevated-Temperature Tensile Properties of Irradiated 2 1/4 Cr-1 Mo Steel", Journal of Nuclear Materials, 126 (1984) 9-17.

3. T. Lauritzen, W. L. Bell, and S. Vaidyanathan, "Effects of Irradiation on the Mechanical Properties of Ferritic Alloys HT-9 and 2.25 Cr-1 Mo," in Proceedings of Topical Conference on Ferritic Alloys for Use in Nuclear Energy Technologies, J. W. Davis and D. J. Michel, eds., TMS-AIME (1984) 623-630.

4. E. P. Simonen, "Calculated Strength of Irradiated Ferritics," Journal of Nuclear Materials, 133 and 134 (1985) 640-643.

5. C. H. Henager, Jr. and E. P. Simonen, "Critical Assessment of Low-Fluence Irradiation Creep Mechanisms," in Effects of Radiation on Materials: Twelfth International Symposium, ASTM STP 870, F. A. Garner and J. S. Perrin, eds., ASTM (1985) 75-97.

6. G. V. Saada, "On the Stability of Quenched Loops in Face-Centered Cubic Metals", Journal of the Physical Society of Japan 18, Supplement III (1963) 41-46.

7. P. Veyssiére and J. Grilhé, "The Influence of Coherent Inclusions on the Nature of Dislocation Loops in Quenched Metals," Philosophical Magazine, 24 (1971) 303-309.

8. J. W. Edington and R. E. Smallman, "Faulted Dislocation Loops in Quenched Aluminum," Philosophical Magazine, 11 (1965) 1109-1123.

9. C. C. Matthai and D. J. Bacon, "On the Unfaulting of Vacancy Loops in BCC Metals," Journal of Nuclear Materials, 125 (1984) 138-151.

10. D. S. Gelles, "A Frank Loop Unfaulting Mechanism in FCC Metals During Neutron Irradiation," in Dislocation Modeling of Physical Systems, M. F. Ashby et al., eds., Pergamon Press (1981) 158-162.

11. N. M. Ghoniem and S. Sharafat," A Numerical Solution to the Fokker-Planck Equation Describing the Evolution of the Interstitial Loop Microstructure During Irradiation," Journal of Nuclear Materials, 92 (1980) 121-135.

12. F. A. Garner and W. G. Wolfer, "A Model for the Evolution of Network Dislocation Density in Irradiated Metals," in Effects of Radiation on Materials: Eleventh International Symposium, ASTM STP 782, H. R. Brager and J. S. Perrin, eds., ASTM (1982) 1073-1087.

13. P. Jung, T. C. Reiley, and W. Kesternich, "Irradiation Creep and Microstructural Changes of Stainless Steel Under Light Ion Irradiation," in Fusion Technology 1980, XI Symposium, Pergamon Press (1981) 1287-1293.

14. H. R. Brager and J. L. Straalsund, "Defect Development in Neutron Irradiated Type 316 Stainless Steel," Journal of Nuclear Materials, 46 (1973) 134-158.

15. W. G. Wolfer and B. B. Glasgow, "Dislocation Evolution in Metals During Irradiation," Acta Metallurgica, 33 (1985) 1997-2004.

MODELING CHROMIUM DEPLETION IN Ni-Cr-Fe ALLOYS

G. S. Was

Department of Nuclear Engineering
Associate Professor of Nuclear Engineering, The University of Michigan
Ann Arbor, Michigan 48109

Abstract

Thermodynamic and kinetic models are constructed to describe the development of the chromium depleted zone in Ni-Cr-Fe alloys heated in the range 773K to 1173K. The thermodynamic model is constructed using the Kohler method for the description of the free energy of a multi-component system. It provides the chromium concentration at the carbide-matrix interface as a function of alloy composition and temperature. The kinetic model tracks the shape of the chromium profile as a function of time at temperature and grain size. These models are formulated into two computer codes, DEPLETE which is applicable to single step isothermal heat treatments and DEPLETE II, which applies to multistep thermal treatments as well as heating and cooling. Model results show that the interfacial chromium concentration decreases for increasing carbon concentration and decreasing heat treatment temperature. Grain shape only becomes important for grain sizes less than 40 μm. Experimental verification of the model is made using high resolution energy dispersive x-ray analysis via STEM. Measured results agree well with model results for the dependence of chromium depletion on various input parameters as well as the magnitude and shape of the chromium depleted zone. Results also show that the maximum temperature attained during the thermal treatment process controls the width of the chromium concentration profile whereas the final temperature controls the grain boundary chromium concentration.

Introduction

For over 15 years, investigators have been developing models of the chromium depletion process in austenitic stainless steel during thermal treatment in the temperature range 773-1173K.[1,2] The motivation is clear. The loss of chromium in a region adjacent to the grain boundary in these alloys can lead to severe intergranular attack in aqueous environments.[3,4] To possess the capability of describing the redistribution of major alloying elements in the vicinity of the grain boundary without continually resorting to laboratory analysis would be an extremely powerful tool. When calibrated against data on intergranular attack, the model not only would become useful in diagnosing potential problems in service components but in preventing them and acting as an alloy design tool to obviate them.

Such a model could also be used to better understand and interpret intergranular corrosion test results. A knowledge of the interrelationships of the major and minor alloying elements would permit an evaluation of the effect of an added impurity (e.g. phosphorus or sulfur) on grain boundary chemistry and subsequent intergranular corrosion and stress corrosion cracking.

However, all models developed to date[1,2] have relied upon the results of bulk corrosion experiments to test the validity of the predictions of chromium depleted zone depth, width and shape. Furthermore, only models for austenitic stainless steel have been developed and the important class of Ni-base alloys, including such technologically important alloys as Inconel 600, have been neglected. What is needed is experimental verification of the chromium-depleted region under the same conditions as those treated in the models. Only then can the validity and accuracy of the models be unambiguously established.

This paper describes an integrated thermodynamic and kinetic model for the redistribution of chromium near the grain boundary of a Ni-16Cr-9Fe alloy (typical of Inconel 600) during thermal treatment. Model results consist of a quantitative description of the chromium depleted zone adjacent to a grain boundary as a function of time at temperature, alloy composition and grain size. The results are compared with chromium depletion profiles measured by quantitative scanning transmission electron microscopy, STEM.

Thermodynamic Model

Assumptions

The following model describes the complete space-time history of the chromium concentration in a sample of a specified composition and grain size, held at temperature for a given length of time. The model applies to a nickel-base alloy containing nominally 16 wt% Cr and 9 wt% Fe with carbon present in the range 0.01 to 0.1 wt%. The alloy must be single phase with all alloying elements uniformly distributed prior to heat treating. The model is based on the following assumptions:

1) The only carbide present is of the form M_7C_3.

2) The metallic content of the carbide is 100% chromium.

3) The carbide forms a continuous film of uniform thickness along the grain boundary.

4) A condition of local equilibrium exists at the carbide-matrix interface.

5) Since the diffusion coefficient of carbon is several orders of magnitude higher than that of chromium, there will at all times be a spatially uniform carbon activity

6) No account is taken of the moving interface between carbide and matrix.

7) No attempt is made to account for carbide nucleation or incubation time.

8) The alloy is treated as a quaternary system in Ni-Cr-Fe-C.

The Ni-Cr-Fe-C system is described by the Kohler model[5] using only binary interaction parameters. According to this model, the free energy of the solution is given by

$$\begin{aligned} G^{fcc} &= x_1 G^{fcc}_{Cr} + x_2 G^{fcc}_{Ni} + x_3 G^{fcc}_{Fe} + x_4 G^{fcc}_{C} \\ &+ RT\,[x_1 \ln x_1 + x_2 \ln x_2 + x_3 \ln x_3 + x_4 \ln x_4] \\ &+ \frac{x_1 x_2}{x_1+x_2}\{x_1 ACRNI + x_2 ANICR\} + \frac{x_1 x_3}{x_1+x_3}\{x_1 ACRFE + x_3 AFECR\} \\ &+ \frac{x_1 x_4}{x_1+x_4}\{x_1 ACRCC + x_4 ACCCR\} + \frac{x_2 x_3}{x_2+x_3}\{x_2 ANIFE + x_3 AFENI\} \\ &+ \frac{x_2 x_4}{x_2+x_4}\{x_2 ANICC + x_4 ACCNI\} + \frac{x_3 x_4}{x_3+x_4}\{x_3 AFECC + x_4 ACCFE\}, \end{aligned} \quad (1)$$

where G^{fcc}_j is the free energy of the pure component in the fcc structure relative to its reference state in cal/mole,

AXXXX are the binary interaction parameters given in Table I,

and Cr = x_1, Ni = x_2, Fe = x_3, and C = x_4, where the x_i are atom fractions in the fcc phase, or generalized,

$$G^{fcc} = \sum_i x_i G_i + RT \sum_i x_i \ln x_i + \sum_{ij} \frac{x_i x_j}{x_i + x_j}\{x_i g_{ij} + x_j h_{ji}\}. \quad (2)$$

The partial molal free energy of the ith component is given by

$$\overline{G}_i = RT \ln x_i + G_i + \sum_{ij} \frac{x_i x_j}{x_i + x_j} \left[\frac{x_j}{x_i + x_j} + 1 - x_i\right] g_{ij}$$

$$+ \sum_{ij} x_i \left[\left(\frac{x_j}{x_i + x_j}\right)^2 - \frac{x_i x_j}{x_i + x_j}\right] h_{ij}$$

$$- \sum_{jl} \{g_{il} \left[\frac{x_j^2 x_l}{x_j + x_l}\right] + g_{lj} \left[\frac{x_l^2 x_j}{x_j + x_l}\right]\}$$

$$\text{and} \quad i \neq j; \quad i \neq l; \quad j \neq l. \tag{3}$$

In particular, for chromium and carbon we have,

$$\overline{G}_{Cr}^{fcc} = G_{Cr}^{fcc} + RT \ln x_1$$

$$+ \frac{x_1 x_2}{x_1 + x_2} \{\frac{x_2}{x_1 + x_2} + 1 - x_1\} ACRNI + x_2 \{(\frac{x_2}{x_1 + x_2})^2 - \frac{x_2 x_1}{x_1 + x_2}\} ANICR$$

$$+ \frac{x_1 x_3}{x_1 + x_3} \{\frac{x_3}{x_1 + x_3} + 1 - x_1\} ACRFE + x_3 \{(\frac{x_3}{x_1 + x_3})^2 - \frac{x_3 x_1}{x_1 + x_3}\} AFECR$$

$$+ \frac{x_1 x_4}{x_1 + x_4} \{\frac{x_4}{x_1 + x_4} + 1 - x_1\} ACRCC + x_4 \{(\frac{x_4}{x_1 + x_4})^2 - \frac{x_4 x_1}{x_1 + x_4}\} ACCCR$$

$$- \frac{x_2^2 x_3}{x_2 + x_3} ANIFE - \frac{x_3^2 x_2}{x_2 + x_3} AFENI$$

$$- \frac{x_2^2 x_4}{x_2 + x_4} ANICC - \frac{x_4^2 x_2}{x_2 + x_4} ACCNI$$

$$- \frac{x_3^2 x_4}{x_3 + x_4} AFECC - \frac{x_4^2 x_3}{x_3 + x_4} ACCFE, \tag{4}$$

and

$$\overline{G}_C^{fcc} = G_C^{fcc} + RT \ln x_4$$

$$+ \frac{x_4 x_1}{x_4+x_1}\{\frac{x_1}{x_4+x_1} + 1 - x_4\}\,ACCCR + x_1\{(\frac{x_1}{x_1+x_4})^2 - \frac{x_1 x_4}{x_1+x_4}\}\,ACRCC$$

$$+ \frac{x_4 x_2}{x_4+x_2}\{\frac{x_2}{x_4+x_2} + 1 - x_4\}\,ACCNI + x_2\{(\frac{x_2}{x_2+x_4})^2 - \frac{x_2 x_4}{x_2+x_4}\}\,ANICC$$

$$+ \frac{x_4 x_3}{x_4+x_3}\{\frac{x_3}{x_4+x_3} + 1 - x_4\}\,ACCFE + x_3\{(\frac{x_3}{x_3+x_4})^2 - \frac{x_3 x_4}{x_3+x_4}\}\,AFECC$$

$$- \frac{x_1^2 x_2}{x_1+x_2}\,ACRNI - \frac{x_2^2 x_1}{x_1+x_2}\,ANICR$$

$$- \frac{x_1^2 x_3}{x_1+x_3}\,ACRFE - \frac{x_3^2 x_1}{x_1+x_3}\,AFECR$$

$$- \frac{x_2^2 x_3}{x_2+x_3}\,ANIFE - \frac{x_3^2 x_2}{x_2+x_3}\,AFENI. \qquad (5)$$

Table I Binary Interaction Parameters and ΔG_C^{FCC-Gr}, $\Delta G_{Cr}^{FCC-BCC}$ in cal/mole with T in K

Parameter	Ref.
$ACRNI = -2000+1.1202\times10^{-3}T^2-1.8649\times10^{-6}T^3$	14
$ACRFE = 1770-1.5T$	14
$ACRCC = -52000$	15
$ANICR = -6000+2.2651\times10^{-3}T^2-6.231\times10^{-7}T^3$	14
$ANIFE = -8320+5.8327\times10^{-3}T^2-2.4859\times10^{-6}T^3$	14
$ANICC = -1900+1.03T$	14
$AFECR = 1770-1.5T$	14
$AFENI = 500-9.1573\times10^{-4}T^2+3.9029\times10^{-7}T^3$	14
$AFECC = -22600-0.7T$	14
$ACCCR = -52000$	15
$ACCNI = -2600+16.18T$	14
$ACCFE = -37100+1.39T$	14
$\Delta G_C^{FCC-Gr} = 33100-3.5T$	13
$\Delta G_{Cr}^{FCC-BCC} = 2500+0.15T$	14

The condition of local equilibrium at the carbide-matrix interface is given by

$$K = \frac{a_{Cr_7C_3}}{(a_{Cr})^7 (a_C)^3}, \tag{6}$$

where K is the equilibrium constant and $a_{Cr_7C_3}$ is the activity of the carbide which is taken to be unity. The equilibrium constant, K, is defined as

$$\Delta G^o_{Cr_7C_3} = -RT \ln K, \tag{7}$$

where $\Delta G^o_{Cr_7C_3}$ is the Gibbs free energy of formation of Cr_7C_3. The data of Chase et al.[6] are used in this study.

In addition, since the carbide is taken to be composed solely of chromium and carbon, the ratio of nickel to iron is everywhere constant,

$$x_2/x_3 = \text{constant}, \tag{8}$$

and

$$x_1 + x_2 + x_3 + x_4 = 1. \tag{9}$$

To completely characterize a quaternary system one needs, in addition to the binary terms, ternary and quaternary terms. However, the use of such an extensive description of this system is pointless unless the thermodynamic data exists to permit assignment of values to these terms. Because of the lack of experimental data and tabulated ternary interaction parameters for this system, a single ternary interaction parameter will be introduced. Following the Kohler method[7], the quaternary system will be approximated as a ternary system of Cr + (Ni+Fe) + C. The additional term in the expression for the free energy of the system is

$$\Delta G^{term} = x_1 x_2 x_4 \text{ AT}, \tag{10}$$

where x_1 = Cr
x_2 = Ni+Fe
x_4 = C
and AT = the temperature dependent ternary interaction parameter in cal/mole.

Since a single ternary term is used to describe a quaternary system, the components x_2 and x_3 are lumped together and are treated as a single component. Therefore, the partial molal free energies corresponding to those in eqn. (10) are

$$\overline{G}_1 = -x_1 x_2 x_4 \text{ AT} ,$$

and

$$\overline{G}_4 = x_1 x_2 (1-x_4) \text{ AT} . \tag{11}$$

The parameter AT becomes the sole parameter with which the total thermodynamic model can be calibrated. It is not, however, a free parameter but is determined using the solubility of carbon in Ni-Cr-Fe as follows. For a given temperature and alloy composition, the matrix carbon level, x^{γ}_C, is set at the solubility limit. A value of AT is selected and x^i_C and x^i_{Cr} are calculated (according to the solution procedure in section 2.2) using eqns (4-6, 8 and 9) and compared with x_C^γ and x_{Cr}^γ respectively. The value of AT is iteratively adjusted until $x_C^\gamma = x_C$ and $x_{Cr}^\gamma = x_{Cr}$. If this process is carried out in the range 573 K to 1173 K, AT is determined to vary linearly with temperature (in K) according to

$$AT = -28.39\ T + 48258 \quad \text{cal/mole.} \tag{12}$$

This is based on solubility data taken from ref. 8,

$$\ln x_C = 9.0 - 14713/T, \tag{13}$$

where x_C is in wt% and T is in K.

Solution Procedure

The objective is to solve for the chromium and carbon concentrations at the carbide-matrix interface. As a first step, eqns. (4) and (5) are rewritten as

$$RT \ln a_{Cr} = G_{Cr}^{fcc} + RT \ln x_1 + \text{etc...}, \tag{4a}$$

and

$$RT \ln a_C = G_C^{fcc} + RT \ln x_4 + \text{etc....} \ . \tag{5a}$$

Equations (4a), (5a), (6), (8) and (9) comprise a system of 5 equations in 5 unknowns, x_1, x_2, x_3, x_4, and a_{Cr}. Recall that a_C is known from the matrix concentration. The solution procedure is as follows:

1) a_C is calculated from equation (5a) given the matrix composition.

2) a_{Cr} is calculated from equation (6) and (7). This is the value of the chromium activity at the carbide-matrix interface.

3) Equations (4a), (5a), (8) and (9) are then solved simultaneously for the concentrations of the alloying elements at the interface. In practice, x_2 and x_3 are eliminated from equations (4a) and (5a) using equations (8) and (9) resulting in a system of two non-linear equations in two unknowns, x_1 and x_4.

4) A completely discretized Newton-Raphson[9] method is used to solve for the chromium and carbon levels.

This is accomplished as follows. If we define a function $f_i(x_1,x_2,\ldots,x_n)$ such that $f_i(x_1,x_2,\ldots,x_n) = 0$ for $i = 1,2,\ldots n$, then Newton Raphson's method (Newton's method) for one dimension (n=1) is

$$f'(x_k)(x_{k+1}-x_k) + f(x_k) = 0 \qquad k=1,2,\ldots,n \ . \tag{14}$$

Generalizing to n dimensions, Newton-Raphson's method becomes

$$\underline{\underline{f}}'(\underline{x}^{(k)})(\underline{x}^{(k+1)}-\underline{x}^{(k)}) + \underline{\underline{f}}(\underline{x}^{(k)}) = 0 \ , \tag{15}$$

where $\underline{\underline{f}}'(\underline{x}^{(k)})$ is an n x n matrix with elements

$$f'_{ij}(\underline{x}) = \frac{\partial f_i}{\partial x_j}(\underline{x}), \qquad 1\leq i,\ j\leq n, \tag{16}$$

and $(\underline{x}^{(k+1)} - \underline{x}^{(k)})$ and $\underline{f}(\underline{x}^{(k)})$ are column vectors. In our case, n=2 and

$$f_1(x_1,x_4) = G_{Cr}^{fcc} + RT\ \ln x_1 - RT\ \ln a_{Cr} + \text{etc...} = 0$$

$$f_2(x_1,x_4) = G_{C}^{fcc} + RT\ \ln x_4 - RT\ \ln a_{C} + \text{etc...} = 0 \tag{17}$$

are complicated functions of x_1 and x_4. Therefore, the partial deviatives in eqn (16) are estimated by use of a difference approximation,

$$\frac{\partial f_i(\underline{x})}{\partial x_j} \simeq \frac{\underline{f}_i(\underline{x}+\underline{h}) - \underline{f}_i(\underline{x})}{h}, \tag{18}$$

where $\underline{h} = 0.001\underline{x}$ has been found to produce satisfactory results. The method of solution is, therefore, a fully discretized Newton-Raphson method. Convergence is usually achieved in 1-3 iterations.

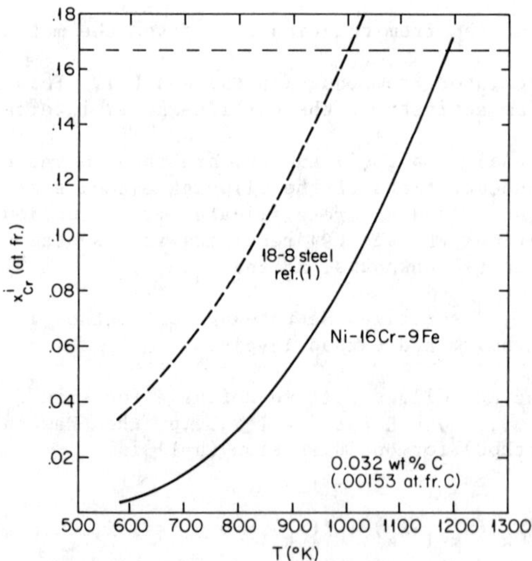

Fig. 1 Equilibrium chromium concentration at the carbide-matrix interface as a function of temperature.

The paraequilibrium chromium concentration at the grain boundary as a function of temperature is given in Fig. 1. Note that for this alloy, the minimum grain boundary concentration of chromium is 7.55 at% or about 6.8 wt%. This agrees well with STEM results[10] of Cr levels in the range 6-8 wt%

Cr for short thermal treatment times at 973 K. Further, the grain boundary Cr concentration reaches the solubility limit at 1180 K in agreement with solubility data. At low temperatures, the model predicts lower Cr concentrations, as expected. Thus. inclusion of a single term produces a match between the free energy expression and the solubility data, resulting in a thermodynamic model that produces interfacial concentrations of chromium and carbon that are in reasonable agreement with existing STEM data. Next, a kinetic model must be developed to describe the space-time evolution of the depleted zone.

Kinetic Model

Determination of the spatial profile of chromium at any time during thermal treatment requires solution of the time-dependent concentration equation

$$\frac{\partial C}{\partial t} = D_{Cr} \nabla_r^2 x(r,t). \tag{19}$$

In the following model, eqn. (19) is discretized and solved numerically,

$$V_n \frac{x_n^i - x_n^{i-1}}{\Delta t} = -D_{Cr}\left(A_n \frac{x_n^i - x_{n-1}^i}{\Delta r_{n-1}} - A_{n+1} \frac{x_{n+1}^i - x_n^i}{\Delta r_n}\right), \tag{20}$$

where x_j^i = atom fraction of chromium at node j and time i,
Δr_j = width of the jth annulus,
A_j = surface area of the jth shell,
V_j = volume of the jth annulus,
D_{Cr} = diffusion coefficient of Cr in 76Ni-15Cr-9Fe (from ref. 11),
Δt = time increment.

Rearranging gives

$$\frac{-DA_n \Delta t}{V_n \Delta r_{n-1}} x_{n-1}^i + \{1 + \frac{D\Delta t}{V_n}(\frac{A_n}{\Delta r_{n-1}} + \frac{A_{n+1}}{\Delta r_n})\}x_n^i - \frac{DA_{n+1}\Delta t}{V_n \Delta r_n} x_{n+1}^i = x_n^{i-1} . \tag{21}$$

Note that the chromium concentration at the interface, x_{Cr}^i, will be grain size dependent since $h = \sum_{j=1}^{N} r_j$. For a sphere consisting of N shells, we have N equations in N+2 unknowns. The left side of each equation contains the Cr concentration for the present time step and the RHS is the Cr concentration at the previous time step. The system of equations is tridiagonal except for the first and last equations which contain one extra term each. The first term of the first equation is eliminated by setting $x_0 = x_1$ since J = 0 at the center of the sphere. The last term on the RHS of the last equation is eliminated by readjusting the right hand side of the last equation to include the x_{n+1} term which is the interfacial chromium concentration for the current time step. The N equations now form a tridiagonal matrix

$$
\begin{aligned}
d_1x_1 + f_1x_2 \qquad\qquad\qquad\qquad &= c_1, \\
e_2x_1 + d_2x_2 + f_2x_3 \qquad\qquad\qquad &= c_2, \\
e_3x_2 + d_3x_3 + f_3x_4 \qquad\qquad &= c_3,
\end{aligned}
\tag{22}
$$

$$
\begin{aligned}
e_{n-1}x_{n-2} + d_{n-1}x_{n-1} + f_{n-1}x_n &= c_{n-1}, \\
e_nx_{n-1} + d_nx_n &= c_n - f_nx_{n+1},
\end{aligned}
$$

where the values of x_j are at the time step i and the c_j are at the time step i-1, and the d_j, e_j and f_js correspond to the coefficients of the x^i_j in eqn (21). This system of equations can be solved as follows using a Crout reduction algorithm.(12)

$$x_n = c'_n, \quad x_j = c'_j - f'_jx_{j+1} \qquad (j=n-1,n-2,\ldots,1),$$

where

$$c'_1 = \frac{f_1}{d_1}, \quad c'_j = \frac{c_j - e'_jc'_{j-1}}{d_j} \qquad (j=2,3,\ldots,n)$$

$$f'_j = \frac{f_j}{d_j} \qquad (j=1,2,\ldots,n-1),$$

$$d'_1 = d_1, \quad d'_j = d_j - e'_jf'_{j-1} \qquad (j=2,3\ldots,n),$$

and

$$e'_j = e_j \qquad (j=2,3,\ldots,n).$$

Code Structure

Two versions of the code have been constructed according to the complexity of the thermal treatment history. For single step isothermal anneals, the code version named DEPLETE is most appropriate. For multistep thermal treatments or the simulation of heating or cooling, DEPLETE II is used. The basic difference between the two codes is in the solution procedure logic. The codes function as follows.

At the beginning of a time step, the kinetic model receives, as input from the thermodynamic model, the chromium and carbon concentrations at the interface for the specified alloy composition and thermal treatment temperature. In DEPLETE, the progression in time is really controlled by fixing the amount of carbon and chromium that flow to the carbide and solving for the time it takes to complete this process. This approach is followed because the precipitation process terminates not after a given amount of time, but when the matrix carbon level reaches the solubility limit. Since the flow rate is so strongly dependent on temperature, the time step size can vary by orders of magnitude and create real difficulty in developing a universal time stepping algorithm.

Each time step, the matrix carbon level is decremented toward the solubility limit. Given the chromium diffusion coefficient, the interfacial chromium concentration, x^i_{Cr}, and the chromium concentration at the last node, x^N_{Cr}, the time to diffuse a corresponding amount of chromium (equal to 7/3 of the carbon increment) into the grain boundary is calculated according to

$$\Delta t = \frac{\Delta Cr}{AD_{Cr}} \frac{\Delta r}{\Delta x} \tag{22}$$

where $\frac{\Delta x}{\Delta r}$ is defined as $\frac{x^N_{Cr}-x^i_{Cr}}{r_N-r_i}$ and Cr is the amount of Cr diffused into the grain boundary in time Δt. Using this time and the chromium concentration profile from the previous time step, the set of N equations is solved. The solution produces a new value of the chromium concentration at the last node, x^N_{Cr}, which is then used with D_{Cr} and x^i_{Cr} to arrive at a new estimate of the time to diffuse the same amount of chromium into the grain boundary. The iteration process is a predictor-corrector scheme where

$$\Delta t^{new} = \Delta t^{old} \times \frac{(x^N_{Cr} - x^i_{Cr})^{old}}{(x^N_{Cr} - x^i_{Cr})^{new}} . \tag{23}$$

During an isothermal anneal, the slope of the chromium concentration profile at the grain boundary is a monotonically decreasing function of time. Therefore estimates of the flow of Cr and C into the grain boundary, based on the Cr concentration profile at the previous time step, will tend to underpredict the time needed to deposit a given carbon increment to the boundary. Hence the new time step size is increased by the ratio of the slopes of the Cr profiles at the grain boundary between successive iterations. This iteration continues until the change in time between successive iterations falls within a predetermined convergence limit. Approximtely 2-3 iterations are typically needed to meet a convergence limit of 1%. At this point, a new chromium profile is produced and the thermodynamic model produces a new value of x^i_{Cr} and a new increment of carbon for the next time step.

In the case of a multistep time-temperature history, the precipitation process is followed by the use of constant time steps. Here the parameter used to determine convergence is the amount of C or Cr that flows to the grain boundary. Solving eqn (22) for ΔC gives

$$\Delta C = \frac{3}{7} A\, D_{Cr} \Delta t \frac{\Delta x}{\Delta r} , \tag{24}$$

and eqn (23) becomes

$$\Delta C^{new} = \Delta C^{old} \frac{(x_{Cr}^{N} - x_{Cr}^{i})^{new}}{(x_{Cr}^{N} - x_{Cr}^{i})^{old}} . \tag{25}$$

The difference in computational procedure between these two approaches is as follows.

In DEPLETE, the Cr level at the grain boundary is fixed after each C decrement, because thermodynamically, there is a unique value of chromium, x_{Cr}^{i}, in equilibrium with the carbide at the predetermined matrix level of C. Therefore, the iteration over the Cr concentration profile is conducted with the grain boundary value of chromium, x_{Cr}^{i}, remaining fixed.

In the case of DEPLETE II, a fixed time step size produces a value for the amount of Cr or C that flows to the grain boundary in <u>that</u> time increment. The Cr profile, and the value of x_{Cr}^{i}, are updated based on the computed values of ΔC or ΔCr. However, since the slope of the new profile at the grain boundary will differ from that of the old, the amount of C or Cr that diffuses into the boundary in the same Δt will change. Hence, in DEPLETE II, the iteration encompasses both the Cr concentration profile calculation and the thermodynamic calculation for the equilibrium level of Cr at the grain boundary. In either case, only 2-3 iterations over the Cr concentration profile are needed for convergence.

Model Results

The thermodynamic model describing the paraequilibrium level of chromium at the carbide-matrix interface as a function of alloy composition and temperature, and the kinetic model describing the time progression of the chromium profile near a grain boundary are combined into single computer programs. The programs are written in BASIC for execution on an IBM personal computer. The reference conditions for all computed results are given in Table II. To maintain consistency with the thermodynamic model, all results are based on the same alloy composition. This is denoted as the model development alloy in Table II.

Table II Reference Conditions for the Study of Chromium Depletion

		Model Development Alloy	Model Verification Alloy
Heat #:		NX8698	AFR91
Composition: (at.fr)	Ni:	.7379	.73103
	Cr:	.16736	.18038
	Fe:	.09421	.08735
	C:	.00153	.00124
Grain Size:		100 μm	
Thermal treatment:		1373K for 20 min. followed by 973K for 1 to 100 hours.	

The time progression of the chromium profile for the model development alloy is given in Fig. 2. As expected, the grain boundary chromium concentration rises and the width of the depleted zone widens with increasing time. Figure 3 shows the change in the shape of the depleted zone as a function of temperature after heat treating for 100 hours. As expected, for a given thermal treatment time, an increase in temperature causes an increase in x^{1}_{Cr} and the depleted zone width.

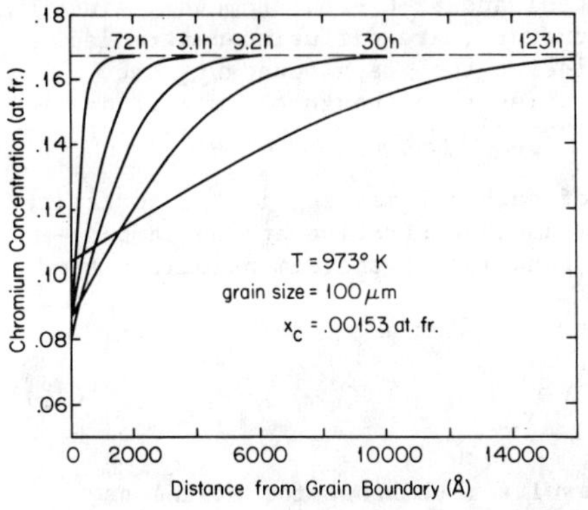

Fig. 2 Time progression of the chromium concentration profile for for the reference conditions.

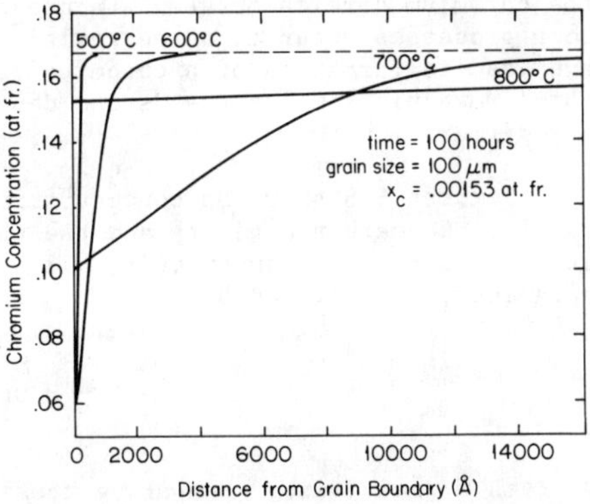

Fig. 3 Effect of temperature on the chromium concentration profile after 100 hours.

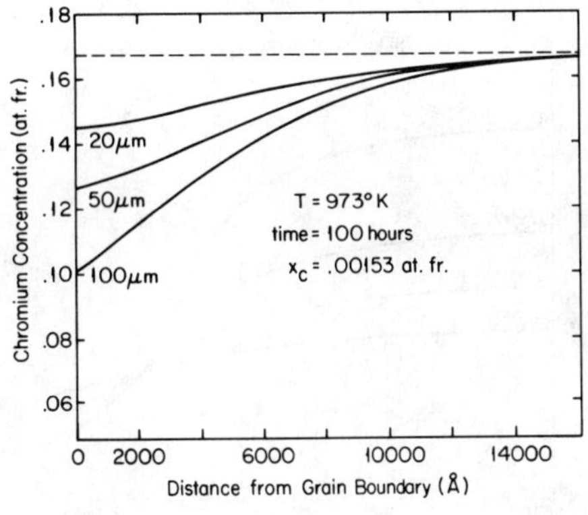

Fig. 4 Effect of grain size on the chromium concentration profile after 100 hours at 973 K.

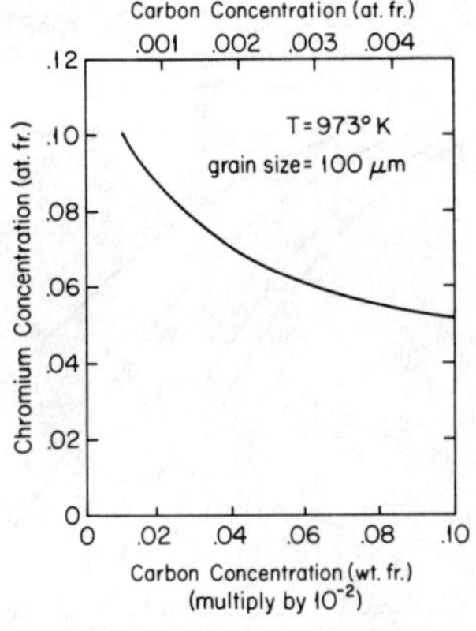

Fig. 5 Equilibrium chromium concentration at the carbide-matrix interface as a function of alloy carbon content.

Figure 4 shows the effect of grain size on the chromium depletion profile after 100 hours at 973K. A smaller grain size means a shorter diffusion path and a more rapid approach to true equilibrium. Thus for the same length of time, a smaller grain material will have a higher grain boundary concentration and consequently a flatter chromium profile at the grain boundary than the larger grain material. The effect of carbon content on x_{Cr}^{i} at time t=0 is shown in Fig. 5. Note that the equilibrium chromium concentration decreases with increasing carbon content and changes most rapidly at low values of carbon.

As a first step in benchmarking this model, a comparison will be made with the results of Stawström and Hillert[1] on stainless steel. There is a difference in base alloy composition and in the structure and composition of the chromium carbide between stainless steel and Ni-Cr-Fe. However, since both processes occur in an austenitic structure, are diffusion-controlled and lead to formation of a chromium carbide at the grain boundary, they should exhibit similar trends and dependencies even though the magnitudes may differ.

Figures 6-8 show the concentration of carbon remaining in the austenite matrix, the carbon activity and the chromium concentration at the carbide-matrix interface, respectively, all as a function of the dimensionless time parameter, τ, defined by

$$\tau = Dt/h^2, \qquad (26)$$

where h = grain size. Comparing these results with those for stainless steel (Figs. 5-7 in ref. 1) it is noted that the curves have the same shape and display the same dependencies. In particular, an intersection of all the curves in the plots of carbon activity vs. τ (Fig. 7) occurs for both Ni-Cr-Fe and stainless steel but at different values of τ and a_C. Referring to Fig. 3, the equilibrium interfacial chromium concentration is plotted as a function of temperature for both Ni-Cr-Fe (solid line) and stainless steel (dashed line). This shows that for an equivalent carbon concentration, the minimum Cr level in Ni-Cr-Fe is considerably less than that for stainless steel.

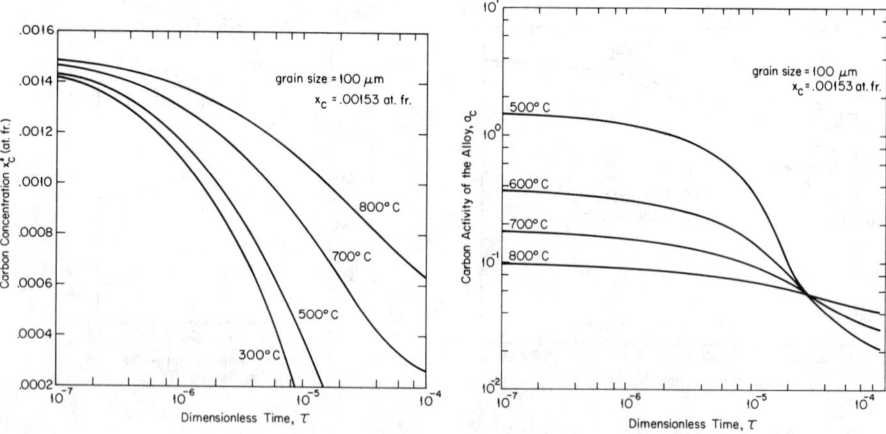

Fig. 6 Carbon content in the matrix as a function of time at temperature.

Fig. 7 Carbon activity of the alloy as a function of time at temperature.

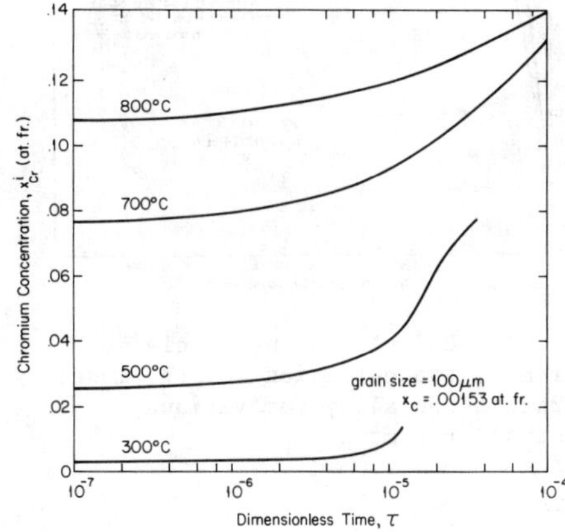

Fig. 8 Chromium concentration at the carbide-matrix interface as a function of time and temperature.

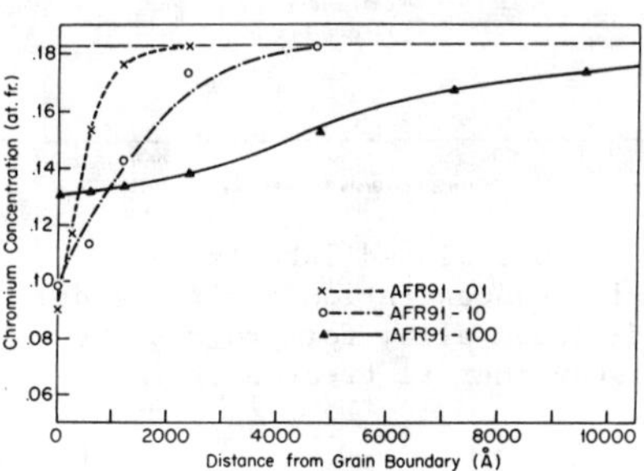

Fig. 9 Chromium concentration profiles after thermal treatments for 1, 10, and 100 hours at 973 K.

Electron beam diffraction analysis of the carbides confirms that they are M_7C_3 (assumption #1) and have a metallic composition of Cr-2%Ni-2%Fe (assumption #2). A comprehensive verification of the model via STEM-EDS using the model verification alloy (Table II) is presented elsewhere.[16] In this verification single step isothermal heat treatments were conducted over the temperature range 873 K-1073 K for 250 hours to 20 minutes respectively. Model results agree well with experimental results as long as intragranular precipitation does not occur, in which case the model predicts deeper Cr depletion profiles. This phenomenon is not modeled in either version of the code and only occurs after long times at temperature. Figure 9 is typical of the agreement between measured and modeled results.

Because DEPLETE II can handle complicated thermal treatment histories, the effect of multiple steps and cooling are shown in Figs. 10 and 11. The steep curve in Fig. 10 results from a single step isothermal anneal at 873 K for 83.3 hours. Note the drastic rise in the grain boundary level and the perturbation near the boundary after application of a 1073 K treatment for only 60 seconds. The high temperature raises the paraequilibrium level of x^i_{Cr} and causes a temporary distortion in the Cr profile until diffusion reëstablishes the quasi-steady state shape with the minimum x_{Cr} at the grain boundary and a flow of Cr and C into the boundary.

An example of a cooling process is shown in Fig. 11. The dashed line represents the Cr concentration profile obtained upon cooling the model verification alloy from 1175 K (above the C solvus) to 933 K in a period of 5 minutes according to an exponential decay. This is compared with the solid curve which is obtained after 5 minutes at 933 K. Note that although the minima are similar, the widths are different. Figures 10 and 11 demonstrate that the maximum temperature obtained during the thermal treatment process controls the width of the Cr concentration profile regardless of when in the process it occurs. However, the Cr level at the grain boundary is determined by the last temperature in the thermal treatment history.

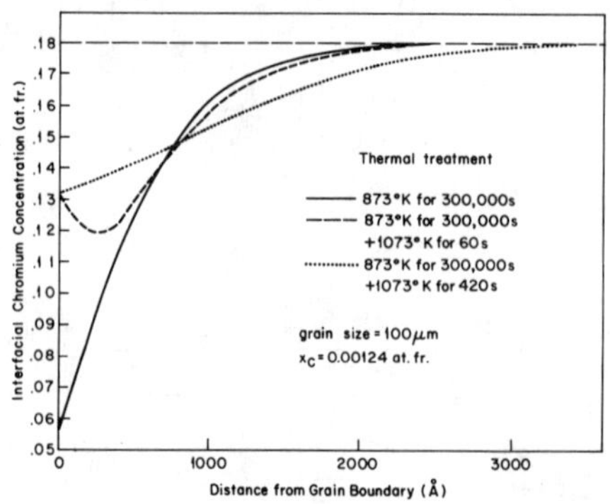

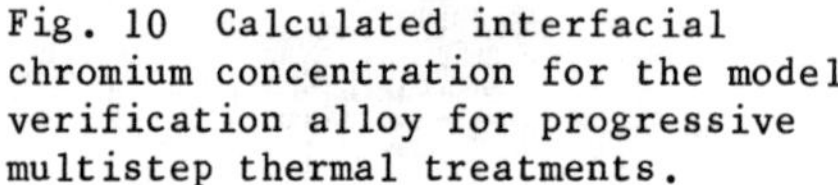

Fig. 10 Calculated interfacial chromium concentration for the model verification alloy for progressive multistep thermal treatments.

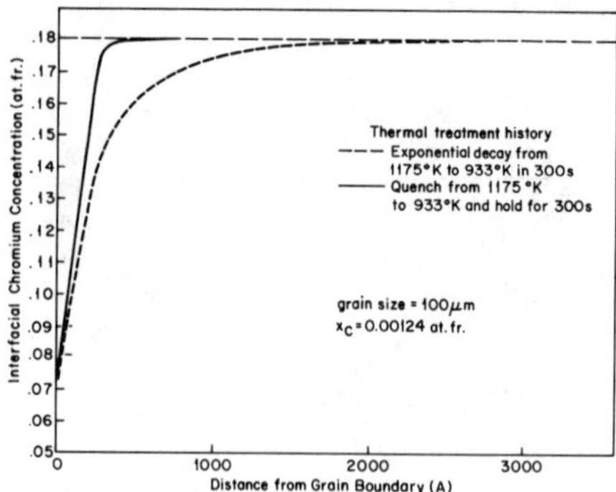

Fig. 11 Calculated interfacial chromium concentration for the model verification alloy for various thermal treatments.

Discussion

Results show that the paraequilibrium level of chromium at the grain boundary is highly sensitive to temperature (Fig. 1) and moderately sensitive to carbon content (Fig. 5). Figure 12 shows the sensitivity of this variable with respect to grain size. For grain sizes $\geq$ 100 μm, the effect of curvature of the grain is minimal. Hence, model results in this regime will be essentailly equivalent to those in which the grain boundary is modeled as an infinite plane. However, below about 40 μm, the rate of change of x^i_{Cr} with grain size becomes significiant. The effect is even more pronounced after shorter times at higher temperatures. Since most engineering alloys have grain sizes in the range 20-40 μm, this becomes an important and practical concern.

In the calculation of the flow of Cr and C to the grain boundary, the amount diffusing into the boundary is proportional to the gradient of the Cr concentration profile at the grain boundary which, numerically, is $\Delta x_n / \Delta r_n$. The value of Δr is a function of r since the spherical grain is divided into equal mass shells. Therefore, the shell boarding the grain boundary, Δr_N, has the smallest thickness. Values of Δr_N = 250 Å are typically used in these calculations, but it is important to know the effect of shell thickness on the calculated value of x^i_{Cr}. Figure 13 shows that x^i_{Cr} is very weakly depenent on Δr_n in the range 50-250 Å. Since computing time is proportioanl to the number of nodes and inversely proportional to the node spacing, a spacing of 250 Å for Δr_N is favored.

One of the most interesting features of the model is the time-dependent nature of the carbide-matrix concentration of chromium which is responsible for some confusion about the shape of the depleted zone after thermal treatment at moderate temperatures for long times. Mulford et al.[17] noted this in stainless steel after 100 hours at 973K and described the chromium profile as Gaussian. Henjered et al.[18] also noticed this in stainless steel after 8 hours at 1023K (which produces a depleted zone of the same width as a thermal treatment of ~50 hours at 973K), but was puzzled by the flat shape at the grain boundary.

The flattened profile near the grain boundary is indeed real and is a consequence of the time-dependent interfacial chromium concentration. As precipitate growth proceeds, the matrix carbon content decreases causing a

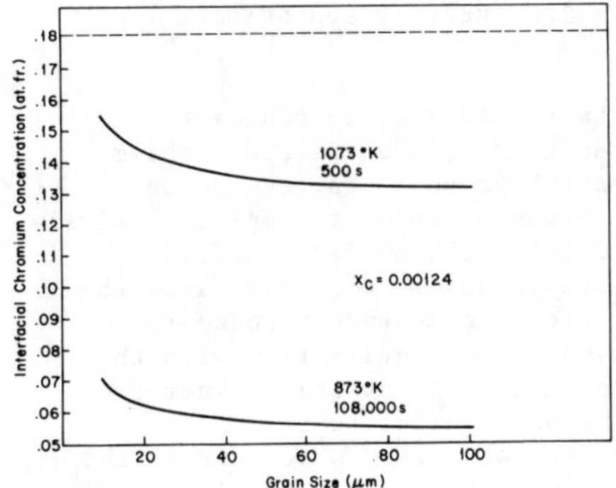

Fig. 12 Calculated interfacial chromium concentration as a function of grain size.

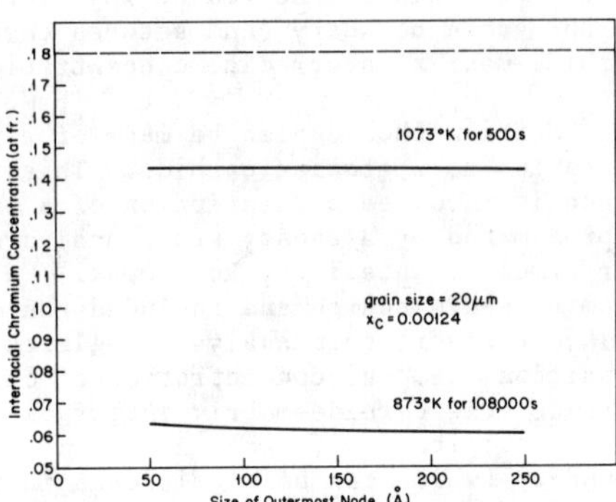

Fig. 13 Calculated interfacial chromium concentration as a function of the size of the smallest node.

decrease in the carbon activity, an increase in the chromium activity (at the carbide-matrix interface) and an increase in the interfacial chromium concentration. The paraequilibrium chromium concentration at the carbide-matrix interface soon rises above the chromium profile near the grain boundary causing a flattening at the grain boundary due to redistribution. This behavior is predicted by the model (Figs. 2-4) and measured in practice (Fig. 9).

The process of chromium depletion and carbide growth modeled in DEPLETE accounts only for volume diffusion. Since grain boundary diffusion is ignored, the question arises concerning the suitability of using the chromium concentration measured at the grain boundary (and between carbides) to represent the carbide-matrix interfacial concentration. The following argument is proposed. The grain boundary diffusion coefficient for chromium is between 5 and 7 orders of magnitude greater than the volume diffusion coefficient at temperatures in the range 1073 to 873K, respectively.[11] Hence, diffusion along the grain boundary is so fast that it is unlikely that a concentration gradient of measurable magnitude will be supported in a grain boundary when the average distance between carbides is less than 5000 Å. Pasparakis et al.[19] have shown that the shape of the concentration contours around a precipitate is a function solely of the homologous temperature. The lower the homologous temperature, the greater the ratio of flux contributed by grain boundary diffusion to that contributed by volume diffusion and the flatter the grain boundary concentration profile of the diffusant. In fact, for temperatures below approximately 0.7 T_m (1173K for AFR91), this ratio asymptotically approaches infinity. Since the maximum temperature used in this study is 1073K (0.64 T_m), we would expect grain boundary diffusion to strongly dominate the flux of solute to the particle and produce a flat solute concentration profile in the grain boundary. Hence, it is the strong domination of solute supply by grain boundary diffusion relative to volume diffusion that gives a flat profile in the grain boundary.

Measurements taken along grain boundary segments at 973 and 1073K show no evidence of a chromium concentration profile in the boundary within the measurement limitations of this technique (±10%). Although grain boundary diffusion is much faster than volume diffusion, it is the supply of solute to the grain boundary which is the rate limiting step. Once at the grain boundary, it moves to the carbide so quickly that no concentration gradient develops. This is the reason why at these temperatures, measurements taken at the grain boundary (and between carbides) accurately approximate the carbide-matrix interfacial concentration.

Mention should also be made of efforts to model grain boundary diffusion to a growing carbide. This is an extremely difficult problem since it involves a description of a transient process that cannot be approximated by a steady state analysis. Perhaps the most complete analysis performed to date is by Rosolowski[20] who modeled the surface diffusion from a needle source and included the effect of volume diffusion into the bulk. However, this analysis neglects 1) the time-dependent boundary condition i.e. the concentration of chromium in paraequilibrium with the carbide (the carbide-matrix interfacial chromium concentration) increases with precipitation (time), 2) volume diffusion parallel to the grain boundary (which can be significant in the case of strong gradients in the grain boundary) and 3) the spherical geometry of the bulk (grain) which is significant for a grain size of 20 μm.

An analysis that includes these effects is extremely complicated, nearly intractable analytically and has not been attempted in this paper. Nevertheless, experimental evidence in combination with the earlier analysis shows that in these experiments, such a detailed treatment of grain boundary diffusion is not necessary to accurately describe the chromium depletion profiles.

Another point of interest is the implication of these results regarding low temperature sensitization (LTS). Inconel 600 is used in steam generator tubing in present day pressurized water reactors. Significant intergranular attack has occured in numerous plants leading to plugging of tubes in most cases[21] and replacement of the entire steam generator in extreme cases.[22] Evidence of chromium depletion in these tubes has been found and concern over in-service sensitization at low temperatures (573-623K) is mounting.[23] However, it is unlikely that a substantial chromium depleted zone can be formed at service temperatures.

Using the volume diffusion coefficient at 573K and assuming volume diffusion to the grain boundary is rate-limiting, then the size of the depleted zone after 40 years is approximately $2(Dt)^{1/2}$ or 3 Å. This figure is substantiated by calculations performed using DEPLETE. Even at 623K the width of the depleted zone would only be 30 Å and at 673K it would be 230 Å. Thus unless another mechanism of diffusion is operative or severe macroscopic intergranular attack can occur with depleted zones of only a few lattice spacings in size, in-service chromium depletion of Inconel 600 in the temperature range 573-623K should not be a problem in the life of the steam generator.

Conclusions

- The chromium depletion code, DEPLETE, models the chromium concentration at the carbide-matrix interface as a function of alloy composition and temperature, and the chromium concentration profile into the matrix as a function of grain size and time at constant temperature. DEPLETE II can handle multistep thermal treatments as well as heating and cooling treatments.

- The thermodynamic model uses binary interaction parameters and a single ternary interaction term with a temperature dependent coefficient determined from the carbon solubility data. Model results are extremely sensitive to the carbon solubility relation.

- The development of the chromium depletion profile with time can be accurately modeled by considering only volume diffusion of chromium to the grain boundary.

- Experimental results show no measurable chromium gradient in the grain boundary, indicating that the grain boundary chromium concentration is probably a good indication of the chromium level at the carbide-matrix interface.

- The level of chromium at the grain boundary in paraequilibrium with the carbide is highly sensitive to temperature, moderately sensitive to carbon content, and sensitive to grain size for values below about 40 μm.

- Results from DEPLETE II show that the maximum temperature attained during the thermal treatment process controls the width of the Cr concentration profile regardless of when in the process it occurs. However, the Cr level at the grain boundary is determined by the final temperature in the thermal treatment history.

- The existence of these codes permits the modeling of Cr concentration profiles that are beyond the measurement capability (resolution) of current STEM-EDS techniques.

Acknowledgement

The author gratefully acknowledges Professor Edward Hucke and Dr. Larry Kaufman for their many stimulating and thought provoking discussions on thermodynamic modeling. This work was supported by the Office of Basic Energy Sciences of the U.S. Department of Energy under grant DE-FG02-85ER45184.

REFERENCES

1. C. Stawström and M. Hillert, JISI, (Jan. 1969) 77.

2. C. S. Tedmon, Jr., D. A. Vermilyea and H. H. Rosolowski, J. Electrochem. Soc., 2 (1971) 192.

3. R. L. Cowan II and C. S. Tedmon, Jr., Advances in Corrosion Science, Vol. 3 (Plenum Press, New York, NY 1973), 293.

4. R. C. Scarberry, S. C. Pearman and J. R. Crum, Corrosion, 32 (1976) 401.

5. F. Kohler, Monatsh. Chemie, 91 (1960) 738.

6. JANAF Thermochemical Tables, J. Phys. Chem. Ref. Data, 4 (1975) 56.

7. M. Hillert, CALPHAD, 4 (1980) 1.

8. Technical service report, "Solid Solubility of Carbon in Inconel Alloy 600", (Huntington Alloy Products Division, Huntington, WV 1971).

9. G. Cahlguist and A. Bjorck, Numerical Methods, (Prentice-Hall, Englewood Cliffs, NJ 1974).

10. G. S. Was, H. H. Tischner and R. M. Latanision, Metall. Trans., 12 (1981) 1397.

11. D. D. Pruthi, M. S. Anand and R. P. Agarwala, J. Nucl. Mater., 64 (1977) 206.

12. F. B. Hildebrand, Methods of Applied Mathematics, 2nd ed., (Prentice Hall, Englewood Cliffs, NJ 1965).

13. L. Kaufman and H. Nesor in Treatise on Solid State Chemistry (edited by N. B. Hanney), Vol. 5, No. 179, (Plenum Press, New York 1975).

14. L. Kaufman and H. Nesor, CALPHAD 4 (1978) 295.

15. L. Kaufman, private communication with author, ManLabs Inc, March 1983.

16. G. S. Was and R. M. Kruger, Acta Metall., 33 (1985) 841.

17. R. A. Mulford, E. L. Hall and C. L. Briant, Corrosion, 39 (1983) 132.

18. A. Henjered, H. Nordin, T. Thorvaldsson and H. O. Andren, Scr. Met., 17 (1983) 1275.

19. A. Pasparakis, D. E. Coates and L. C. Brown, Acta. Met., 21 (1973) 991.

20. J. H. Rosolowski, Metall. Trans., 3 (1972) 285.

21. C. Y. Cheng, "Steam Generator Tube Experience", (Report No. NUREG-0886, Nuclear Regulatory Commission, Washington, D.C. 1982).

22. H. H. Woo and S. C. Lu, "Worldwide Assessment of Steam Generator Problems in Pressurized Water Reactor Nuclear Power Plants", (Lawrence Livermore Laboratory, UCRL-53032 Sept. 1981).

23. E. Serra, "Stress Corrosion Cracking of Alloy 600," (Report No. NP-2114-SR, Electric Power Research Institute, Palo Alto, CA 1981).

COMPUTER MODELING OF THE SOLIDIFICATION OF EUTECTIC ALLOYS:

THE CASE OF CAST IRON

Doru M. Stefanescu, Professor,
and Chandrashekhar Kanetkar, Graduate Assistant
The University of Alabama
Department of Metallurgical Engineering
P.O. Box G
University, Alabama 35486

Abstract

The solidification process of cast iron is controlled, like any other solidification process, by nucleation and growth phenomena. Thus, using nucleation and growth laws, as well as heat transfer laws, the cooling of the simple shape casting has been modeled. We include in our model: the heat released by the casting before solidification, the amount of this heat transferred to the mold (which we calculate in two ways), and the heat released by the casting during solidification. The fraction of solid was calculated using the Johnson-Mehl equation, assuming that nuclei grow as spheres in cast iron (which is correct for both flake and spheriodal graphite iron - eutectic cells), and using three different growth laws to calculate the radius, R, of the eutectic cells. It was thus possible to predict the cooling curve of a simple casting, having a given chemical composition, as a function of the number of nuclei and solidification modulus (size of casting).

1. Introduction

The ultimate goal of modeling microstructural changes during solidification and thermal processing of alloys is to predict the room temperature structure as influenced by the different variables manipulated during cooling of the alloy from the liquid state. The availability of such a computer program will then allow one to predict the mechanical properties based on calculated information on microstructure. Such required information will be, for example for cast iron, as follows: pearlite/ferrite ratio, pearlite fineness, eutectic grain (cell) size, graphite type and size, etc.

Most of the work which is underway at the present time in computer modeling of solidification is based on solving the heat transfer problem. This approach cannot give the required information on microstructure and microstructural changes.

A different approach has been tried, first by Oldfield [1] and then by other investigators [2, 3, 4], using nucleation and growth laws to model the solidification of cast iron.

The model discussed in this paper capitalizes on both methods; that is, it uses a heat transfer approach to model cooling and uses nucleation and growth laws to model structural transformation, as well.

Such a model can be used not only for prediction of structure and mechanical properties, but also to control the melting and liquid treatment processes of cast iron based on calculated information such as the number of eutectic nuclei (inoculation), eutectic undercooling (chill), stable or metastable structure resulting from solidification in various molds (different cooling rates), etc.

The model can also be used to check the validity of existing models for dendritic and eutectic nucleation and growth and to calculate rate values for any type of eutectic alloy.

2. Computer Program

The flow chart of the computer program used (called EUCAST) is given in Figure 1. It covers cooling of cast iron in liquid state, solidification in the two-phase region and in the eutectic interval, cooling of solid cast iron above the eutectoid temperature, eutectoid transformation, and cooling of iron after the eutectoid transformation to room temperature.

Each of these subroutines will be discussed separately.

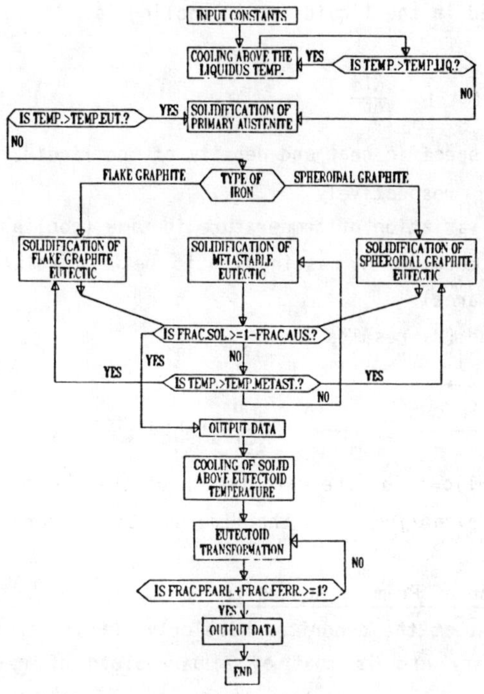

Figure 1. Flow chart of computer program.

2.1. Cooling of Liquid Iron Above the Liquidus Temperature

Both the analytical method and numerical method are used to model the cooling of the casting. In this paper only the analytical method will be discussed.

A number of assumptions were made, as follows: isothermal solidification throughout the casting, semi-infinite mold, cooling starts at pouring temperature.

Then, the heat flow across the metal-mold interface can be written as:

$$\frac{dQ}{dt} = \sqrt{\frac{K_m \rho_m C_m}{\pi t}}\,(T_i - T_o) \tag{1}$$

where K_m: thermal diffusivity of the mold in cal/cm·s·°C

ρ_m: density of the mold in g/cm^3

C_m: specific heat of the mold in cal/g °C

T_i: interface temperature in °C

T_o: ambient temperature in °C

t: time in seconds

The heat evolved in the liquid during cooling is:

$$\frac{dQ}{dt} = MC_L\rho_L \frac{dTi}{dt} \tag{2}$$

where C_L, ρ_L: specific heat and density of the liquid, respectively

dTi/dt: variation of temperature in time (cooling rate)

$M = V/A$: modulus of casting (ratio between volume and surface area)

Equating (1) and (2) results in:

$$dTi = \sqrt{\frac{K_m\rho_mC_m}{\pi}} \; \frac{T_i-T_o}{MC_L\rho_L} \; \frac{dt}{t^{1/2}} \tag{3}$$

This equation allows for the calculation of the change in temperature in the casting while cooling above the liquidus temperature.

2.2. Solidification of Primary Austenite

It was assumed that the dendrites have only primary arms and that the shape of the primary arm is that of a paraboloid of revolution. The dimensions for the model of dendritic growth are shown in Figure 2.

Assuming that there is a fixed correlation between the two axes of this paraboloid, a and b, and taking this correlation to be

$$a = 11.7b^2 \tag{4}$$

as found by Glicksman and Schaefer [6] for pure tin, the volume of one dendrite arm is:

$$V_d = \frac{\pi}{2} b^2a = 0.134a^2 \tag{5}$$

where

$$a_{i+1} = a_i + (da/dt)\Delta t \tag{6}$$

Assuming now that austenite dendrites grow according to a parabolic law, the growth rate of dendrites is:

$$da/dt = \mu_1\Delta T^2 = \frac{\beta \, D_L \, \Delta H_\gamma^2}{4\pi RT_M^3 \, V_m^\gamma \sigma} \; \frac{1+2g^{1/2}}{g} \; \Delta T^2 \tag{7}$$

as shown by Tiller [7], where

β: constant which relates the interface-attachment transition frequency to the diffusional-jump frequency in the liquid.

D_L: self diffusion coefficient in the liquid.

ΔH_γ: latent heat of fusion of austenite.

R: gas constant.

T_M: melting temperature.

v_m^γ: molar volume of austenite.

σ: interface energy.

g: diffuseness parameter.

Assuming, in a first approximation, that $\beta = 1$ and g = 1, the growth constant is $\mu_1 = 3.51 \times 10^{-4}$ cm/s $\cdot$ °K^2.

The number of dendrites per unit area is:

$$N = 1/d \tag{8}$$

where d is the primary interdendritic arm spacing (see Figure 2) and can be calculated as:

$$d = b(G\frac{da}{dt})^{-n} = b(\frac{dT}{dt})^{-0.5} \tag{9}$$

where b: constant = 150 for d in μm (from Figure 5.15, Reference [5])

$G\frac{da}{dt} = \frac{dT}{dt}$: gradient by growth rate product which is the cooling rate

Now the fraction of solid austenite can be calculated as:

$$F_\gamma = N^{3/2} V_d \tag{10}$$

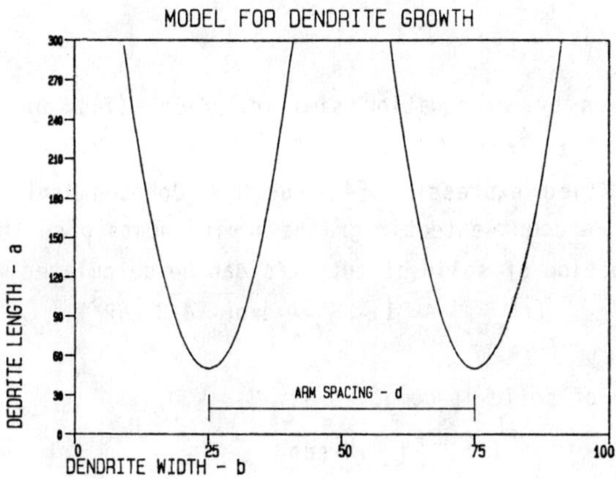

Figure 2. Dimensions for the model of dendritic growth.

The heat flow across the metal-mold interface is given again by Equation (1), while the heat evolved in the liquid during the solidification of primary austenite is:

$$\frac{dQ}{dt} = M\rho_\gamma \quad [C_\gamma F_\gamma + C_L(1-F_\gamma)] \cdot \frac{dTi}{dt} + \Delta H_\gamma \frac{dF_\gamma}{dt} \qquad (11)$$

and equating Equations (1) and (11), the cooling of the liquid-primary austenite mixture is:

$$\frac{dTi}{dt} = [\sqrt{\frac{K_m\rho_m C_m}{\pi t}} \; \frac{Ti-To}{M\rho_s} - \Delta H_\gamma \frac{dF_\gamma}{dt}]/[C_\gamma F_\gamma + C_L(1-F_\gamma)] \qquad (12)$$

2.3. Solidification of the Austenite-Flake Graphite Eutectic

The model for the solidification of the austenite-flake graphite eutectic is based on the following assumptions:

1. The eutectic grains are spherical in shape, with radius R.
2. Each eutectic grain is the product of a nucleation event.
3. Nucleation rate is given by an exponential low, which means that nucleation starts at a given temperature (ΔT critical) with a number of nuclei which is actually a constant N.
4. The eutectic grains start growing at the same time, and grow with the same rate.

Assuming now that growth of the flake graphite eutectic grains occur by a screw dislocation mechanism, the growth rate of the eutectic grains can be calculated as:

$$dR/dt = \mu_2 \cdot \Delta T^2 \qquad (13)$$

where μ_2 is given by an equation similar to Eq. (7), and for $\beta=1$ and g=1 is $\mu_2 = 4.53 \cdot 10^{-4} cm/s \cdot {}^{\circ}K^2$.

Using a modified expression [4] for the Johnson-Mehl [9] equation which takes into account eutectic grains impingements once they touch one another, the fraction of solid of eutectic can be calculated with:

$$F_{eut} = 1 - F_\gamma - \exp(-\ 4/3\ \pi N R^3) \qquad (14)$$

The fraction of solid is computed as:

$$F_{i+1} = F_i + dF_{eut} \qquad (15)$$

where dF_{eut} is calculated from:

$$dF_{eut}/dt = 4\pi(1 - F_{eut} - F_\gamma)R^2 N\ dR/dt \qquad (16)$$

with R computed as:

$$R_{i+1} = R_i + dR \qquad (17)$$

Once F_{eut} is known, cooling in the flake graphite eutectic interval can be calculated with an equation similar to Equation (12).

2.4. Solidification of the Austenite-Spheriodal Graphite Eutectic

Assumptions similar to those for the model of flake graphite eutectic were used for the model of austenite-spheroidal graphite eutectic. It was also assumed that spheroidal graphite nuclei are encapsulated in an austenitic shell right after their formation, further growth occurring by diffusion of carbon through the austenite shell.

Then the growth rate of the austenite shell encapsulating the graphite spheroid can be calculated with the equation derived by Wetterfall et al. [10] for diffusion controlled eutectic growth through the austenite shell:

$$\frac{dR}{dt} = \frac{V_m^G}{V_m^\gamma} D_C^\gamma \frac{1}{R_G(1-R_G/R_\gamma)} \frac{X^{\gamma/L} - X^{\gamma/G}}{X^G - X^{\gamma/G}} \tag{18}$$

where D_C^γ: diffusion coefficient of carbon in austenite.
R_G, R_γ: radius of graphite spheroid and austenite shell.
$X^{\gamma/L}$, $X^{\gamma/G}$: mole fraction of carbon at phase boundaries
X^G: mole fraction of carbon content of graphite

Experimental values seem to indicate that the R_G/R_γ ratio is constant throughout growth, having a value of 0.416 [10]. Using the other values given in Appendix I, Equation (18) becomes:

$$\frac{dR}{dt} = 4.663 \cdot 10^{-10} \frac{\Delta T}{R_G} \tag{19}$$

This equation was then used to calculate F_{eut} and dTi/dt, as shown above for the case of flake graphite eutectic.

2.5. Solidification of the Austenite-Iron Carbide Eutectic

Modeling of the solidification of the iron carbide eutectic seems to be rather complicated because of the complexity of this eutectic. Apparently solidification starts with crystallization of flat plates of Fe_3C, more or less faceted, leading austenite dendrites, with a lamellar eutectic resulting in the direction of the Fe_3C plates. Also, perpendicular to the plates, cooperative growth produces a rod type eutectic Fe_3C-austenite [11, 12]. Growth rates are different in the two growth directions.

At this time a more primitive model was adopted, assuming again that the eutectic grains are spherical in shape rather than elongated.

As solidification of white iron requires considerably higher undercooling than gray iron, it was assumed that growth proceeds by two-dimensional nucleation, and as shown in [7], the growth rate can be calculated as:

$$\frac{dR}{dt} = \mu_4 \exp\left(\frac{\mu_3}{3\Delta T}\right) \tag{20}$$

where

$$\mu 3 = \frac{\pi g V_m^2 \sigma^2}{R a_0^2 \Delta H_{\gamma Fe3c}} \quad \text{in } ^\circ K \tag{21}$$

$$\mu 4 = (2 + g^{-1/2}) \frac{\beta D_L}{a_0} \left(\frac{\Delta H_{\gamma Fe3c}\, \Delta T}{R T_M^2}\right)^{7/6} \quad \text{in cm/s} \tag{22}$$

a_0 = molecular step height normal to the interface in cm

Using data given in Appendix 1, for $\beta = 1$ and $g = 1$, the coefficients were calculated to be: $\mu_3 = 627^\circ K$ and $\mu_4 = 0.908 \ \Delta T^{7/6}$ cm/s

With the growth rate known, the remaining computations are done as for the austenite-flake graphite eutectic.

A parabolic law of the form given in Equation (13) could be also considered for growth. Calculations give a theoretical value $\mu_2 = 1.68 \cdot 10^{-4}$ for $\beta = 1$ and $g = 1$.

2.6. Eutectoid Transformation

In order to produce a model for the eutectoid transformation, it has been assumed that the γ-α transformation proceeds under equilibrium conditions, with austenite partly transforming in pearlite and partly in ferrite.

Data for the nucleation rate of pearlite were taken from Figure 17.15 in Reference [13] and approximated to an exponential law:

$$\frac{dN}{dt} = 1.8 \cdot 10^{22} \exp\left(3.33 \cdot 10^4/T\right) \tag{23}$$

The growth rate of pearlite for 0.78 percent C steel, according to Frye et al. [14], is:

$$\frac{dR}{dt} = 2.86 \ \Delta F \Delta T \exp\left(-24200/RT\right) \tag{24}$$

where ΔF is the free energy of formation of austenite from pearlite, which, using again data from Reference [14], was estimated to be:

$$\Delta F = -1.46\ (T - 996) \tag{25}$$

The volume of pearlite resulting from the decomposition of austenite can now be calculated, using Johnson-Mehl's equation [9], as:

$$F_{perl} = 1 - F_{fer} - \exp[-\frac{\pi}{3}\frac{dN}{dt}(\frac{dR}{dt})^3 t^4] \tag{26}$$

and
$$\frac{dF_{perl}}{dt} = \frac{4}{3}\pi \frac{dN}{dt}(\frac{dR}{dt})^3 [1 - F_{fer} - F_{perl}]\, t^3 \tag{27}$$

To calculate the amount of ferrite resulting from the eutectoid transformation, it was assumed that the transformation is diffusion controlled and occurs at steady state, and that the ferrite grains are spherical in shape and start growing at the same time with the same rate.

The growth rate of ferrite grains has been derived [23] to be:

$$\frac{dR}{dt} = \frac{C_2 - C_3}{C_4 - C_3} \frac{D_C^{\alpha}}{R_{\alpha}^2 (\frac{1}{R_{\alpha}} - \frac{1}{R_0})} \tag{28}$$

where
D_C^{α}: diffusivity of carbon in ferrite
R_{α}: radius of ferrite grain
R_0: radius of ferrite nucleus
C_2: carbon content in the middle of the ferrite grain
C_3: carbon content at the ferrite-austenite interface
C_4: carbon content in the austenite

The number of ferrite nuclei starting to grow is considered equal to the final number of ferrite grains, which for cast iron is between $1.7 \cdot 10^6$ and $2.5 \cdot 10^8/cm^2$.

Now the fraction of ferrite can be calculated as

$$F_{fer\ i+1} = F_{fer\ i} + (\frac{dF_{fer}}{dt})\Delta t \tag{29}$$

where dF_{fer}/dt is calculated with Equation (27).

3. Some Results

3.1. Solidification of the Austenite-Flake Graphite Eutectic (Gray Iron)

The theoretical value of μ_1 in growth equation (7) for austenite has been shown to be $3.51 \cdot 10^{-4} cm/s/{}^{\circ}K^2$, for β=1 and g=1. Other calculations by Subramanian et al. [16] suggest a value of μ_1=0.0016. Running the EUCAST program with the latest value, results in austenite fractions far from the ones which can be calculated from the phase diagram, while for the μ_1 calculated with Equation (7) the computed austenite fractions were very close to the equilibrium ones.

In order to compute the solidification path of gray iron it is necessary to assume a number of nuclei (eutectic grains). Literature values are: 345 to 1000/cm^3 [17], $1.5 \cdot 10^3$/cm^3 [18], $1.7 \cdot 10^4$ to $3 \cdot 10^5$/cm^3 [19]. In our calculations N=1500/cm^3 was used as a base number.

The theoretical value of μ_2 in growth equation (13) for the austenite- flake graphite eutectic, assuming again β=1 and g=1 was shown to be $4.53 \cdot 10^{-4}$cm/s$\cdot$K^2. Other literature values are as follows: $6.4 \cdot 10^{-5}$ [16], $7.25 \cdot 10^{-6}$ to $9.5 \cdot 10^{-6}$ [19], 10^{-7} [20].

An example of the way in which the EUCAST program can be used to determine μ_2 is shown in Figure 3. The upper curve has been calculated using $\mu_2 = 4.53 \cdot 10^{-4}$, which is the theoretical value for β=1 and g=1, while the lower curve was calculated with $\mu_2 = 7.25 \cdot 10^{-6}$.

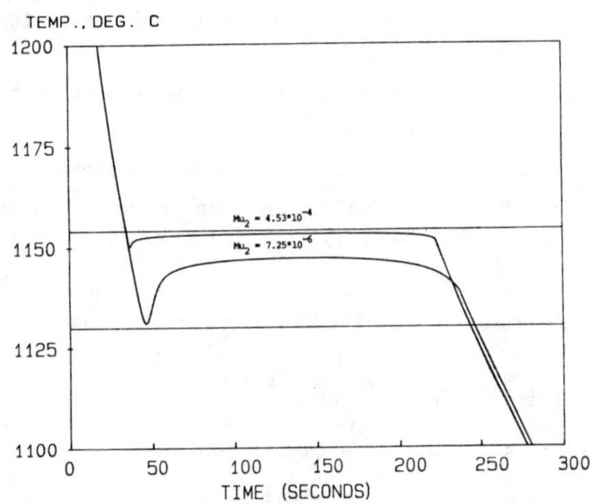

Figure 3. Cooling curves for irons of eutectic composition poured in castings with 0.8 modulus, for different growth rate constants μ_2.

By matching this curve with an experimental curve, it should be possible to determine the correct value for μ_2

The influence of some process variables is shown in Figure 4. It can be seen that, as expected, undercooling is decreased by the increase of either the carbon equivalent or the number of nuclei (inoculation).

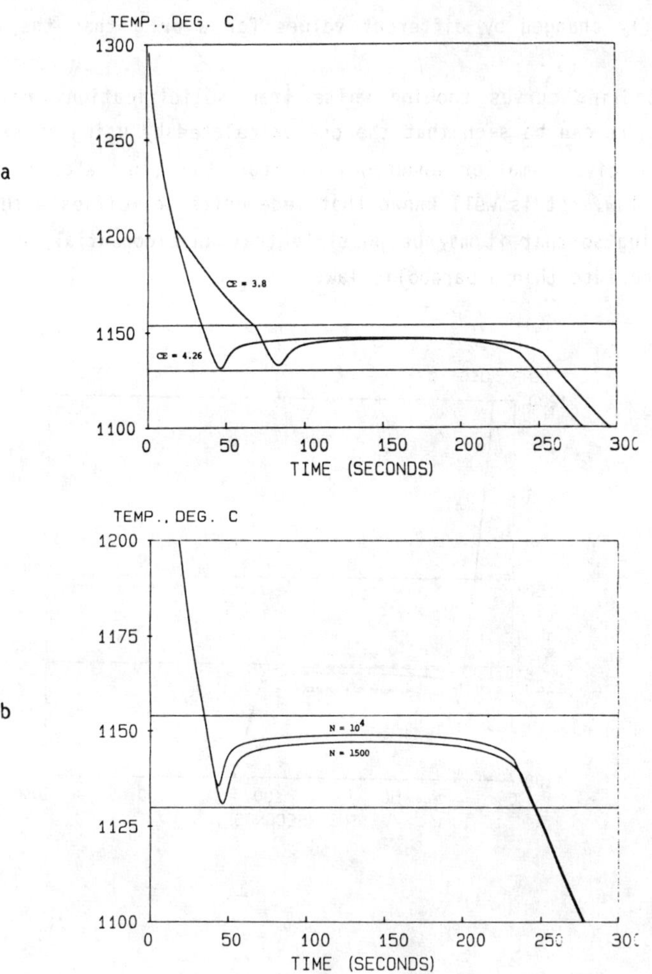

Figure 4. Influence of (a) carbon equivalent and (b) inoculation on the undercooling of the austenite-flake graphite eutectic.

3.2. Solidification of the Austenite-Iron Carbide Eutectic (White Iron)

Assuming $g = 10^{-3}$ and $\beta = 10^{-2}$, the coefficients in the growth law from Equation 20 become $\mu_3 = 0.627$ and $\mu_4 = 6.06 \cdot 10^{-3}$.

Other researchers have assumed parabolic laws of the type given by Equation (13). Some values for the coefficient μ_2 are as follows: 0.001 [16], 0.0025 [22]. Our theoretical value calculated from Equation (7) is $1.58 \cdot 10^{-4}$. This number could of course be dramatically changed by different values for β or g than the unit ones used.

Two cooling curves showing white iron solidification are shown in Figure 5. It can be seen that the one calculated by using an exponential law for R gives smaller undercooling than the one calculated with a parabolic law. It is well known that ledeburite solidifies with very low undercooling so that it may be possible that an exponential law might be more appropriate than a parabolic law.

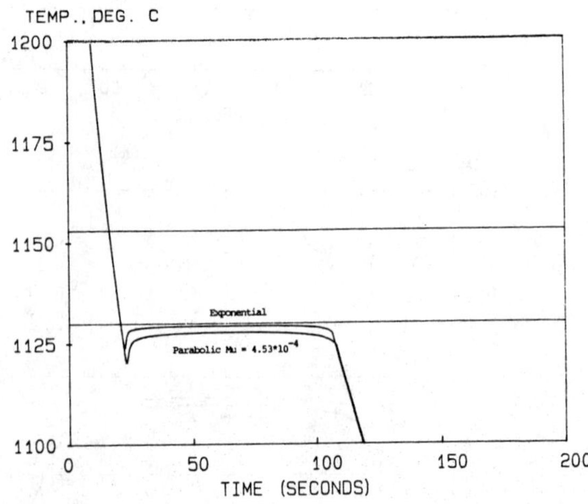

Figure 5. Influence of different growth laws on the calculated cooling curve of eutectic white irons (modulus 0.5).

Figure 6 shows the influence of the modulus of the casting on the solidification of cast iron. As expected, decreasing the modulus results eventually in the solidification of the austenite-iron carbide eutectic (white iron) rather than of the austenite-graphite eutectic (gray iron).

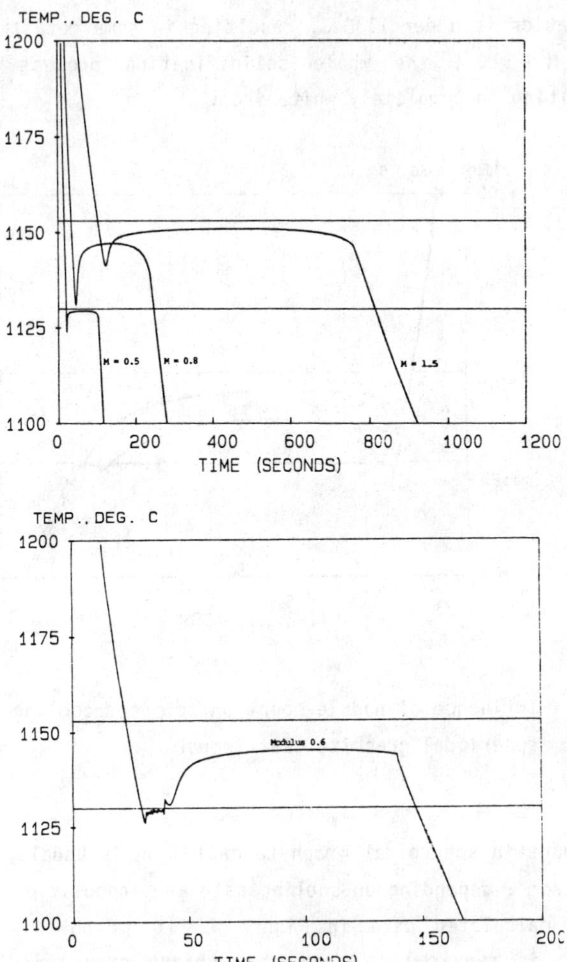

Figure 6. Influence of casting modulus on the gray-white transition for irons of eutectic composition.

3.3. Solidification of the Austenite-Spheriodal Graphite Eutectic (Ductile Iron)

The influence of nodule count on the cooling curves is shown in Figure 7. It can be seen that solidification follows the stable path only at nodule counts higher than $10^{11}/cm^3$. When $N = 10^{10}$, the end of solidification is under 1130°C, resulting in some metastable eutectic, while when $N = 10^9$, the whole solidification process occurs under 1130°C, resulting in completely white iron.

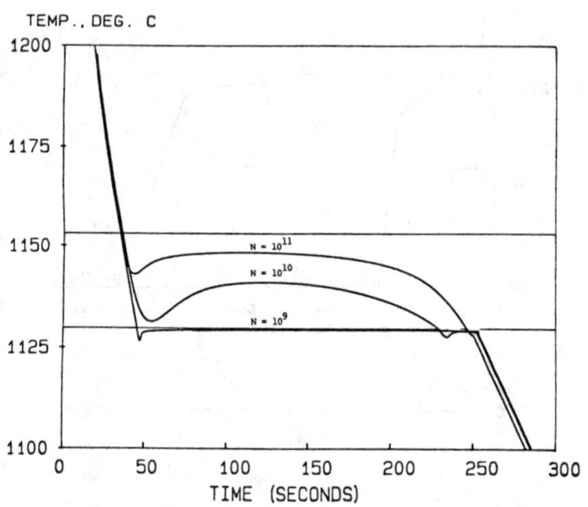

Figure 7. Influence of nodule count on computed cooling curves for spheriodal graphite cast irons.

Nodule count in spheroidal graphite cast iron is usually in the range of 10^6 to $10^7/cm^3$, depending on cooling rate and inoculation.

From the calculated data in Figure 7, it is obvious that a much higher number is required in order to achieve gray solidification, if solidification is to occur only by diffusion of carbon through the austenite shell. It is probably safe to conclude from this analysis that solidification of the austenite-spheroidal graphite eutectic proceeds at least during the first part by independent growth of the eutectic phases, with possible later encapsulation of graphite in austenite.

3.4. Eutectoid Transformation

The results of computations for the eutectoid transformation are shown in Figures 8 and 9. It was assumed that $R_\alpha = 11\mu m$ and $R_o = 10\mu m$. It can be seen that, as expected, the amount of ferrite increases for lower cooling rates (higher modulus).

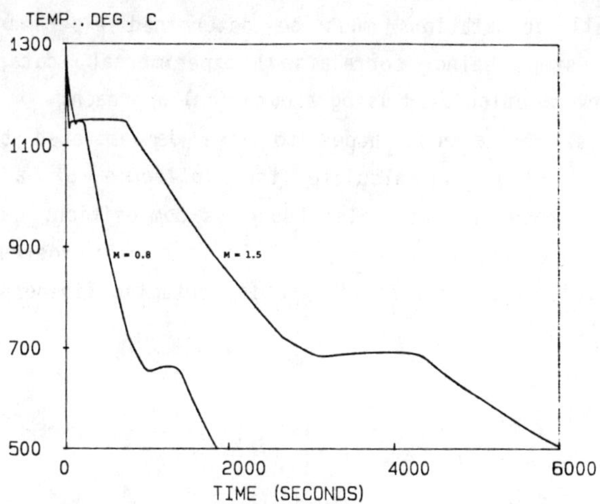

Figure 8. Influence of casting modulus on the eutectoid transformation.

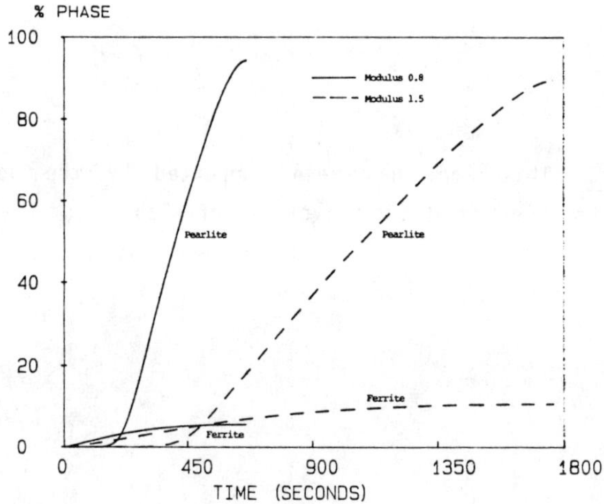

Figure 9. Variation of pearlite and ferrite content during the eutectoid transformation for castings with different modulus.

Concluding Remarks

This paper must be regarded as the first one of many steps to be taken toward the completion of the EUCAST program. None of the above results should be considered final. It is quite obvious that the validity of all computations must be determined experimentally, the logical next step being correlating experimental data with the theoretical curves, calculated using a numerical approach.

Nevertheless, the authors hoped to have demonstrated that such a program can be used to calculate the influence of a number of experimental or production variables (such as composition, casting size, mold materials, pouring temperature, inoculation) on microstructure (gray to white transition, pearlite/ferrite ratio, eutectic fineness, eutectoid fineness, etc.).

Acknowledgment. This work has been supported by the Department of Metallurgical Engineering at the University of Alabama.

REFERENCES

1. W. Oldfield, Trans. ASM, 59 (1966) 945.

2. D.M. Stefanescu, S. Trufinescu, Z. Metallkde, 65 (1974) 610.

3. K.C. Su, I. Ohnaka, I. Yamauchi, T. Fukusako, The Physical Metallurgy of Cast Iron, Ed. H. Frederiksson and M. Hillert, North-Holland (1985) 181.

4. H. Frederiksson and I. Svenson, The Physical Metallurgy of Cast Iron, Ed. H. Frederiksson and M. Hillert, North-Holland (1985) 273.

5. M. Flemings, Solidification Processing, McGraw-Hill (1974) 364 p.

6. M.E. Glicksman, R.J. Schaefer, Acta Met., 14 (1966) 1126.

7. W.A. Tiller, Solidification, ASM, (1971) 59.

8. D.M. Stefanescu, The Physical Metallurgy of Cast Iron, Ed. H. Frederiksson and M. Hillert, North-Holland (1985) 151.

9. W.A. Johnson, R.F. Mehl, Trans. AIME, 135 (1939) 416.

10. S.E. Wetterfall, H. Fredriksson, M. Hillert, J.I.S.I., (1972) 323.

11. M. Hillert, M. Steinhauser, Jernkontorets Ann., 144 (1960) 520.

12. M. Hillert, V.V. Subba Rao, The Solidificaion of Metals, Iron Steel Inst. Publ. 110, London (1968).

13. R.E. Reed-Hill, Physical Metallurgy Principles, Brooks/Cole (1973) 923 p.

14. J. H. Frye, Jr., E.E. Stansbury, D.L. McElroy, Trans. AIME (1953) 219.

15. C. Zener, Trans. AIME 167 (1946) 550.

16. S.V. Subramanian, D.A.R. Kay, G.R. Purdy, Trans. AFS (1982) 589.

17. W. Oldfield, Recent Research on Cast Iron, Ed. H.D. Merchant, Gordon and Breach (1968) 347.

18. D.M. Stefanescu, F. Martinez, I.G. Chen, Trans. AFS (1983) 205.

19. M. Fredriksson, S.E. Wetterfall, The Metallurgy of Cast Iron, Ed. B. Lux, I. Minkoff, F. Mollard, Georgi Publishing Company, Switzerland, (1975) 277.

20. G. Lesoult, M. Turpin, The Metallurgy of Cast Iron, Ed. B. Lux, I. Minkoff, F. Mollard, Georgi Publishing Company, Switzerland, (1975) 255.

21. J.W. Cahn, W.B. Hillig, G.W. Sears, Acta Met. 12 (1964) 1421.

22. M. Hillert, Recent Research on Cast Iron, Ed. H.D. Merchant, Gordon and Breach (1969) 101.

23. D.M. Stefanescu, C. Kanetkar, Unpublished work at The University of Alabama (1985).

APPENDIX 1: CONSTANTS

Thermophysical properties of sand molds: Ref. [5]

K_m = $14.5 \cdot 10^{-4}$ cal/cm·°C·s

ρ_m = 1.5 g/cm^3

C_m = 0.27 cal/g·°C

Constants for calculation of coefficients in growth laws:

D_L = $5 \cdot 10^{-5}$cm^2/s

D_C^γ = $9 \cdot 10^{-7}$cm^2/s Ref. [4]

ΔH_γ = 62.86 cal/g = 3481 cal/mol

$\Delta H_{\gamma G}$ = 55.7 cal/g = 3111 cal/mol Ref. [8]

$\Delta H_{\gamma Fe_3C}$ = 50.9 cal/g = 2843 cal/mol Ref. [8]

ΔH_{perl} = 20.5 cal/g Ref. [15]

v_m^γ = 7.1 cm^3/mole

$v_m^{\gamma G}$ = $v_m^{\gamma Fe_3C}$ = 7.98 cm^3/mole Ref. [8]

v_m^G = 5.34 cm^3/mole

σ austenite = 204 erg/cm^2 = $4.88 \cdot 10^{-6}$ cal/cm^2

σ cast iron = 191 erg/cm^2 = $4.6 \cdot 10^{-6}$ cal/cm^2

a_o = $3.44 \cdot 10^{-8}$ cm Ref. [4]

$X^{\gamma/L} - X^{\gamma/G}$ = $3.66 \cdot 10^{-4}\,\Delta T$ Ref. [4]

$X^G - X^{\gamma/G}$ = 0.0909 Ref. [4]

A MICROSTRUCTURAL MODEL

FOR ANNEALING OF Al-KILLED STEEL

Hanne Mathiesen

CENTER FOR INDUSTRIAL RESEARCH (SI)
P.O. Box 350 Blindern
0314 Oslo 3, Norway

Abstract

A quantitative, isothermal model for the recrystallization and the precipitation and growth of aluminium nitrides in Al-killed steel is derived. The model is based on available theories for the individual metallurgical reactions. The time and temperature dependence in the theoretical equations are determined by regression analysis of data both from the literature and from this investigation. The expressions obtained are combined into a model for isothermal annealing, which takes account of the interaction between the individual reactions.

The model which at this stage is limited to a constant chemical composition and a given cold-rolling reduction will later be generalized. The complete model should be utilized for killed steels with variable chemical compositions and processes which involve variable rolling reductions and general heating cycles.

Introduction

The mechanical properties of Al-killed steel depend strongly on the grain structure of the sheet. Whether for instance the recrystallized grains become either equiaxed or elongated is determined by the chemical composition, coiling temperature after hot-rolling and heating rate in batch annealing. The elongated grain structure, which seems to be a good indication of a favourable texture for deep drawing, is obtained when recrystallization interacts strongly with the precipitation of aluminium nitrides. It is believed that AlN-particles or pre-precipitation clusters of Al and N decrease the nucleation rate and hinder the movement of recrystallized boundaries, and thereby control the grain structure and the texture (1) and (2).

In order to obtain an optimal quality of the semi-finished steel sheets, it is most important to control the annealing cycle carefully. This investigation was undertaken to develop a quantitative model for the annealing of Al-killed steel. The model is divided into two parts, one part describes the precipitation and growth of AlN. The other part deals with recrystallization and takes into account the effect from second phase particles and segregations on the recrystallization kinetics. The first part serves as an input to the recrystallization model.

As a first stage, the model is developed for isothermal annealing. This gives the opportunity to base the model on available theories for the relevant metallurgical reactions, recrystallization, precipitation and growth of second phase particles. The model is preliminary limited to a given chemical composition and a given cold-rolling reduction.

In order to make a model for industrial annealing of Al-killed steel, the quantitative expression derived in the present model must be integrated over the actual thermal cycle. Integrating in that way allows one to model both the precipitation and the growth of AlN that occur during slow cooling of the hot-rolled coil and the interaction between recrystallization and precipitation during batch annealing.

Theory

Recrystallization

Recrystallization kinetics is most commonly described by an Avrami relation

$$X = 1 - \exp(-Bt^k) \quad , \quad B = B_o \exp(-\frac{Q_{rex}}{RT}) \tag{1}$$

where X is the volume fraction transformed, t is the isothermal annealing time and T the annealing temperature. B_o and k are numerical constants, R is the gas constant and Q_{rex} is the activation energy for recrystallization.

Equation (1) is empirical and yields little information about the fundamental processes involved in recrystallization. The macroscopic quantities B and k can be described by microscopic parameters for the recrystallization process such as nucleation and growth rate (3), but the correlation between microscopic and macroscopic quantities is generally rather complex.

Both second phase particles and impurity atoms segregated to the grain boundaries are known to reduce the velocity of the moving boundaries and, therefore, slow down the transformation markedly. From this consideration it is clear that B_o and k are not constant but depend on the microstructure. Our model allows, therefore, these quantities to vary during recrystallization.

Precipitation

Precipitation from a supersaturated solid solution is controlled by the nucleation and growth of the new phase. The nucleation is expected to occur at a constant rate during the whole transformation so that a wide range of particle sizes exist at any time. The growth of the particles measured by the geometrical mean particle radius, r_t, in a log-normal distribution will normally follow a growth law

$$r_t = r_o + b(Dt)^m \quad , \quad D = D_o \exp(-\frac{Q_{diff}}{Rt}) \qquad (2)$$

D is the diffusion coefficient, b is a numerical constant and m is a parameter dependent on the reaction mechanism. r_o is the mean particle size at t = 0.

The overall kinetics of precipitation can be described by the following relation between annealing time and nitrogen in solid solution

$$C_t = C_\infty + (C_o - C_\infty) \exp(-At^n) \quad , \quad A = A_o \exp(-\frac{Q_{presp}}{RT}) \qquad (3)$$

where C_o and C_∞ are the initial and equilibrium concentrations of nitrogen in solution, and C_t is this concentration at time t. A_o is a constant and Q_{presp} is the activation energy for precipitation which is generally the sum of the activation energies for nucleation and growth.

Segregation

Due to a positive interaction force between solutes and faults (4), the concentration of foreign atoms in the grain boundary, C_{bound}, will be increased over the mean concentration, C_{matrix}

$$C_{bound} = C_{matrix} \exp(\frac{U_B}{RT}) \qquad (4)$$

U_B is the energy of interaction between impurity atoms and the grain boundary. As a first approximation, it is assumed that $U_B >> RT$ so that nearly all impurity atoms will tend to segregate the boundaries. The degree of segregation C_{seg} in a given time can then be estimated from the diffusion distance;

$$C_{seg} = C_o \sqrt{D\,t} \quad , \quad D = D_o \exp(-\frac{Q_{diff}}{RT}) \qquad (5)$$

D is the diffusion coefficient in ferrite and C_o is the matrix concentration of nitrogen at t = 0.

Material and Method

A quantitative model for precipitation of AlN and recrystallization is developed from the equations discussed above. All the equations are transformed to linear relationships between functions of the microstructural parameter and functions of annealing time and temperature. The parameters in Equations (1), (2), (3) and (5) are then estimated by multivariable least-squares method.

The input data is as far as possible taken from the literature. Supplementary experiments were carried out when necessary for a complete description.

Material

The material used for the experiments is hot-rolled, commercial, low-carbon, Al-killed steel. Chemical composition of this steel, No. 1, and other steels from the literature are given in Table 1.

Table I. Chemical Composition of the Steels investigated in wt%

Steel No.	C	Mn	S	P	Al	N	Ref.
1	0.08	0.32	0.004	0.016	0.046*	0.007	
2	0.04	0.28	0.023	0.015	0.041*	0.0053	1
3	0.05	0.41	-	-	0.040	0.0045*	5

* Acid soluble

Experimental Method

Samples from the hot-rolled strip were exposed to several heat-treatments in order to simulate different coiling temperatures and to obtain samples with different particle distributions prior to recrystallization. The specimens were annealed isothermally at 720 ^{0}C and 620 ^{0}C for periods up to 4 hours. After cold-rolling to 50% reduction, the samples were again isothermally annealed at 500 ^{0}C and 600 ^{0}C for recrystallization.

Recrystallization was followed by hardness and texture measurements in order to separate recovery and recrystallization. The fall in hardness before any detectable change of the texture of the sheet was attributed to recovery. After recrystallization had started, the loss in hardness was taken to be proportional to the volume fraction transformed.

Transmission electron microscopy was used to measure the size of the precipitated AlN particles in order to analyse growth kinetics both in the hot-rolled strip and in cold-rolled material.

Due to the fact that the AlN-particles are very small, below about 25 nm in this investigation, and that they have weak contrast in TEM, it is difficult to image the precipitates in a deformed structure. For this reason the growth of particles in the deformed structure was estimated by measuring the particle size before cold-rolling and in the just recrystallized structure.

Results and Discussion

Precipitation and growth of AlN

Reaction Mechanisms. Fig. 1 shows the precipitation of aluminium nitride by means of nitrogen combined as AlN. The solubility of nitrogen in equilibrium with AlN in ferrite is considered to be zero (6). When precipitation follows Equation (3), the data for each temperature should describe a straight line of slope k in Fig. 1. From the existing data for precipitation in a deformed matrix, the best estimate for n is 0.84 $\pm$ 0,15, but there is considerable scatter in the experimental data. The plot for precipitation in undeformed material, however, exhibits a different time dependence. In the beginning of the precipitation sequence m is 0.87 ± 0.08, the same as obtained for precipitation in a deformed matrix. After 60-70% precipitation, the reaction is restrained markedly and n is reduced to 0.10 ± 0.05.

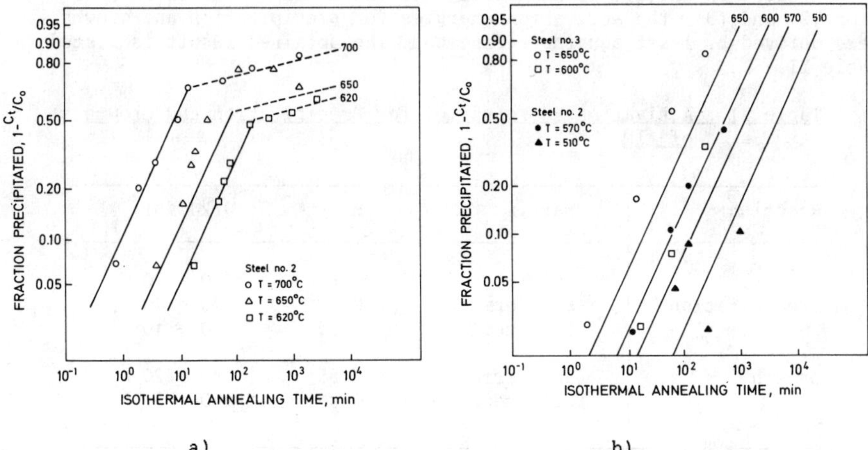

a) b)

Figure 1 - Fraction nitrogen precipitated as aluminium nitrides. The curves are calculated from Equation (6) and (7).
a) undeformed material
b) deformed material

Growth of the particles are shown in Fig. 2. Growth in undeformed material yields a m-value of 0.83 ± 0.10 in Equation (3), which is consistent with the obtained m-values for precipitation. In the absence of complete experimental data for growth in the deformed matrix, m is assumed to be 0.85, because the growth and precipitation are controlled by the same mechanism.

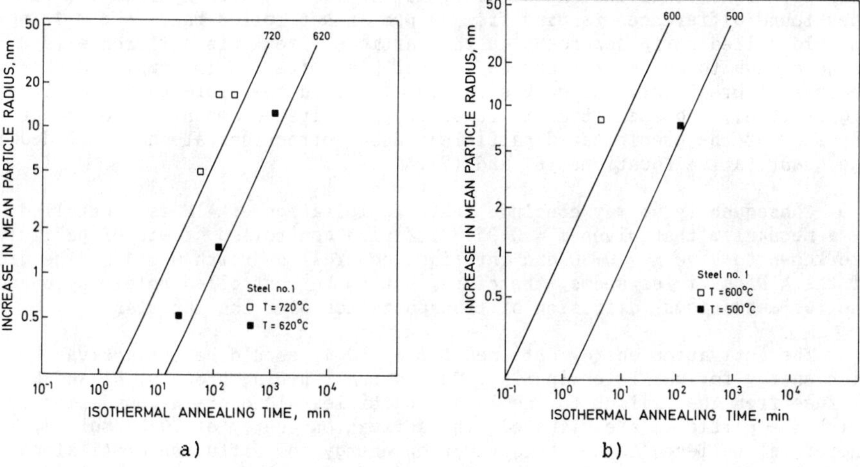

a) b)

Figure 2 - Growth of aluminium nitrides. The curves are calculated from Equation (8) and (9).
a) undeformed material
b) deformed material

In the following it is therefore supposed that m = n = 0.85 (weighted mean value) for precipitation and growth of AlN, except for the regions above the break of the curves in Fig. 1 where m = 0.1, see discussion below.

Activation energy. With knowledge of the m- and n-values in Equation (2) and (3), the activation energies for precipitation and growth were derived by least squares refinement. The obtained result is listed in Table II.

Table II. Activation Energies (Q) for Precipitation and Growth of AlN

Reaction	Matrix	m	Q(kJ/mol)
Precipitation	undeformed	0.85	240 ± 20
	undeformed	0.10	70 ± 10
	deformed	0.85	110 ± 10
Growth	undeformed	0.85	140 ± 20
	deformed	0.85	160 ± 30

Discussion

The reduction in reaction rate that is observed for precipitation of AlN in undeformed material is not believed to be a real effect, but a consequence of the use of Beeghly's method of analysis. When m changes from 0.85 to 0.10, the mean particle size is about 5 nm, calculated from Equation (8). It is, therefore, reasonable to believe that the precipitation is complete at this point, and the slope of the latter part of the curve only results from the fact that more and more particles attain the critical size for detection, $r \simeq 10$ nm. Research done by Hoogovens Group BV, (7), shows that the amount of N in solid solution determined by Beeghly's method is markedly higher than that measured by hot extraction with hydrogen, (8). They found differences ranging from 29 ppm in hot-rolled band to 2 - 12 ppm in cold-rolled strip dependent on the particle size. This difference is large enough to adjust the curves so that precipitation is completed where the curves break. All the curves in Fig. 1 should therefore be raised to a higher level, and the actual increase in precipitated amount is dependent on the size of the precipitated particles. Such corrections are not included in the quantitative Equations (6) and (7).

Consequently we may conclude that precipitation of AlN is controlled by a mechanism that gives m = 0.85. Diffusion controlled growth of particles are known to give m = 0.50 and interface controlled growth m = 1.0. The growth of the AlN particles seems, therefore, not to be controlled solely by one of the two mechanisms, diffusion or transportation over the interfase.

The activation energy obtained for m = 0.1, should be the activation energy for particle ripening. During the ripening process, Al and N diffuse from the smaller to the larger particles along grain boundaries on which the particles are situated. The activation energy of 70 kJ/mol is, therefore, believed to be the activation energy for diffusion of Al along grain boundaries since Al is the slower diffusing element.

Before precipitation is completed, the rate determined process is diffusion of Al from the nearly dislocation-free areas in the grain - or sub-grain interior. The activation energy for growth both in the cold rolled and soft condition is then expected to equal the activation energy for the diffusion of Al in ferrite.

The activation energy for precipitation is found to be (110 $\pm$ 10) kJ/mol in deformed condition, whereas the energy necessary for growth is about (150 $\pm$ 30) kJ/mol. Precipitation consists of both nucleation and growth of particles, so Q_{presp} should in any case not be less than Q_{growth}. To get a consistent model it is therefore assumed that Q_{presp} should equal Q_{growth} in the cold-rolled steel. This means that nucleation is a far more easy process than diffusion in a deformed structure, and the activation energy for nucleation is vanishing small compared to growth.

The best obtainable estimate for Q_{growth} is (120 $\pm$ 20) kJ/mol, weighted mean of 160, 140 and 110 kJ/mol. The only reported value for activation energy for diffusion of aluminium that is found is 185 kJ/mol, (9). This value is considerably higher than what is obtained here. We shall not pay too much attention to the difference, because the reference gives no information about the crystal structure and temperature range.

Quantitative description

On the basis of the previous discussion of m- and n- values and activation energies the following quantitative relationships were derived.

Precipitation of AlN in undeformed material

$$\ln^2 \frac{C_o}{C_t} = (27 \pm 2) + (0.85 \pm 0.1)\ln t - \frac{(240 \pm 20)\ \text{kJ/mol}}{RT} \quad (6)$$

Precipitation of AlN in deformed material, 50% cold reduction

$$\ln^2 \frac{C_o}{C_t} = (10.8 \pm 0.1) + (0.85 \pm 0.1)\ln t - \frac{(120 \pm 20)\ \text{kJ/mol}}{RT} \quad (7)$$

Growth of AlN in undeformed material

$$\ln(r-r_o) = (12.3 \pm 0.1) + (0.85 \pm 0.1)\ln t - \frac{(120 \pm 20)\ \text{kJ/mol}}{RT} \quad (8)$$

Growth of AlN in deformed material, 50% cold rolled reduction

$$\ln(r-r_o) = (16.4 \pm 0.4) + (0.85 \pm 0.1)\ln t - \frac{(120 \pm 20)\ \text{kJ/mol}}{RT} \quad (9)$$

t is the annealing time in minutes and T the temperature in Kelvin. These relationships are shown i Figs. 1 and 2 together with the experimental data.

Recrystallization

When recrystallization is described by Equation (1), the data for recrystallized fraction should fall on a straight line with slope k when $\ln^2 (\frac{1}{1-X})$ is plotted against lnt.

Some of the data in Fig. 3 falls on straight lines with different inclinations. Other data describes curved lines which means that k decreases during the transformation. It is believed that these variations in k are a consequence of segregation and precipitation.

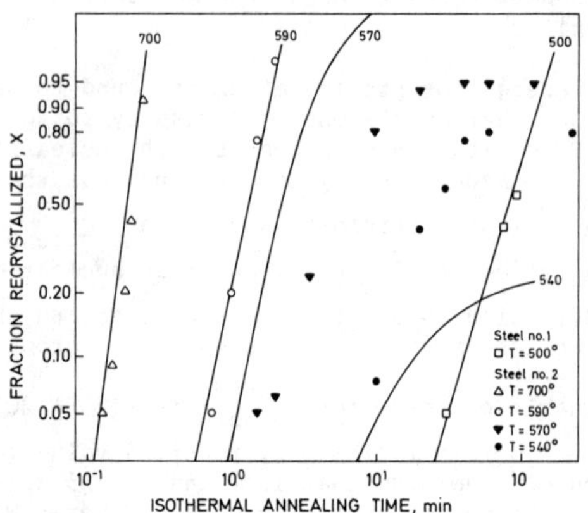

Figure 3 - Recrystallization of samples with different microstructure. The curves are calculated from Equations (12) and (13).

k-value. The influence of nitrogen segregated to boundaries was analysed in samples with almost identical particle distribution and essential no precipitation during recrystillization.

An empirical relationship is derived between k and the particle-distribution and the degree of segregation in the system. The precipitation and segregation model serves as input for the calculation of k.

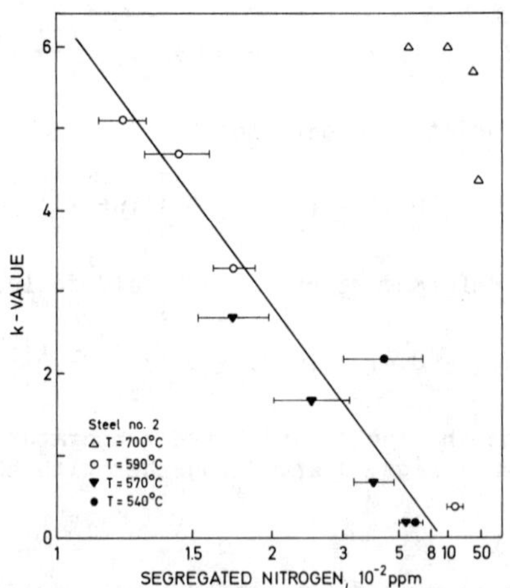

Figure 4 - Effect of nitrogen segregation on k in Equation (1).

k appeared to be a nearly linear function of $1/C_{seg}$, in the range of observation, Fig. 4.

$$k = -0.7 + 5.5 \cdot 10^{-3} \frac{1}{\sqrt{D_N t}\, C_o} \tag{10}$$

where t is the annealing time in seconds, C_o is the matrix concentration of nitrogen in ppm and D_N the diffusion coefficient for nitrogen in ferrite; $D_o = 3.0 \cdot 10^{-3}\ cm^2s^{-1}$ and Q = 76 kJ/mol (10). Since $k < 0$ has no physical interpretation, k should be taken as zero when Equation (10) gives k less than zero.

The k's obtained for transformation at 700 ^{0}C do not seem to follow this law.

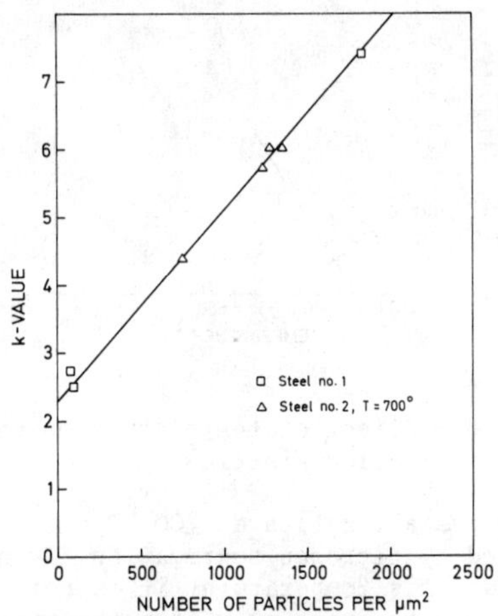

Figure 5 - Effect of aluminium nitrides on k in Equation (1).

The effect from second-phase particles was investigated in samples with negligible influence from segregated nitrogen; i.e. samples with almost all nitrogen precipitated or samples annealed at 700 ^{0}C. k showed good linearity with the number of particles per unit area, Fig. 5.

$$k = 2.3 + 2.8 \cdot 10^3 \frac{f}{r^2} \qquad (11)$$

The volume fraction of AlN-particles, f, and the particle-size in nm are calculated from the precipitation and growth model, Equations (6)-(9).

<u>Activation energy</u>. With a known k-value, calculated from the expressions obtained above, the activation energy for recrystallization is found by least squares refinement of Equation (1).

To minimize the effect of inaccuracies in the calculated k-values which will be integrated during the recrystallization, only data from the very beginning of the transformation was used. As Fig. 6 shows, the data gives a well-defined activation energy for recrystallization of (650 ± 50) kJ/mol.

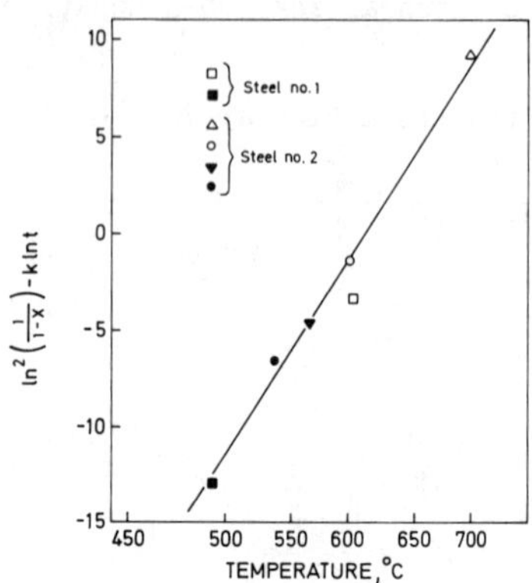

Figure 6 - Effect of temperature on the recrystallization kinetics.

Discussion. For recrystallization at 700 °C the k-value does not seem to be markedly influenced by nitrogen segregation. Even though the transformation rate is high at this temperature, it is believed that the moving grain boundaries are unable to break away from the nitrogen atmosphere. The inconsistency at 700 °C may be a consequence of the temperature dependence in Equation (4). This model is based on the assumption that all free nitrogen atoms will segregate to the boundaries independent of the temperature. From Equation (4), however, it is clear that the tendency to segregation will decrease with increasing temperature, and the error caused by this fact will be more serious at higher temperatures. It is, therefore, believed that the error in the calculated degree of segregation become considerable at 700 °C, and the real segregation at this temperature should be much lower.

In further work, the calculation of C_{seg} should take account of the temperature dependent equilibrium distribution between segregated and free impurity atoms in Equation (4).

Since second-phase particles are known to inhibit the movement of grain boundaries (Zener drag), the recrystallization rate should decrease with increasing Zener parameter, $\frac{f}{r}$. For the data obtained here, however, the k-value is greater at larger values of $\frac{f}{r}$ and $\frac{f}{r^2}$. This might be due to the fact that the recovery process is slowed down by the second-phase particles. The second-phase particles inhibit sub-grain formation and growth during recovery so that the dislocation density during recrystallization will be greater at larger particle densities. The driving force for recrystallization is thereby increased with $\frac{f}{r^2}$ resulting in a higher k-value.

At this stage the model has been developed for the effect of segregated N-atoms and second-phase particles independently. In most cases, however, these effects will be combined. The effect of impurity segregation is known to be dependent of the velocity of the grain boundaries (11), and is therefore not independent of the particle distribution. Eventhough, the two effects are assumed independent and additive, this should not lead to difficulties, because when the particle effect dominates, the effect from segregated nitrogen should be less important, and vica versa.

The combined effect from precipitated AlN and N segregation are then expressed as a sum:

$$k = -4.4 + 5.5 \cdot 10^{-3} \frac{1}{\sqrt{D_N\, t}\; C_o} + 2.8 \cdot 10^{3} \frac{f}{r^2} \qquad (12)$$

Recrystallization involves rearrangement of dislocations to form a recrystallization front and subsequent migration of the front. The activation energy which is the sum of the activation energy for the two processes involved, should always be greater than the activation energy for self-diffusion, 240 kJ/mol (10). The difference between the obtained activation energy, (650 ± 50) kJ/mol, and 240 kJ/mol should then be the energy associated with nucleation of new grains.

Quantitative description. From the results obtained above, recrystallization kinetic in Al-killed steel may be described by the following relation:

$$\ln^2 \left(\frac{1}{1-X}\right) = (90 \pm 10) + k \ln t - \frac{(650 \pm 50)\ \text{kJ/mol}}{RT} \qquad (13)$$

where k represents the interaction with the segregation and precipitation of AlN, and k is given by Equation (12). t and T are annealing time in minutes and temperature in Kelvin.

Fig. 3 shows recrystallization curves calculated on from Equation (12) and (13). The straight lines in the diagram correspond well with the experimental data. The curved lines, however, deviate to some degree from the data. This might be due to a relative small inaccuracy in the calculated k-values. Since k gives the inclination of the curve, such errors will be integrated and cause large deviation at the latter stage of the recrystallization process.

Conclusion. The quantitative model developed for isothermal annealing of Al-killed steel, Equation (6)-(9), (12) and (13), corresponds reasonable with the experimental data. The recrystallization process is strongly retard by nitrogen segregation at the grain boundaries and precipitated aluminium nitrides. When recrystallization kinetics is described by an Avrami relation, the order of the reaction, k in Equation (1), accounts for this interaction.

To make the model applicable to industrial annealing, the isothermal expression should be integrated over the desired thermal cycle.

References

1. R.H. Goodenow, "Recrystallization and Grain Structure in Rimmed and Aluminium-Killed Low-Carbon Steel", Transactions of the ASM, 59 (1966) 804-823.

2. J.T. Michalak and R.D. Schoone, "Recrystallization and Texture Development in a Low-Carbon, Aluminium-Killed Steel", Trans. met. of AIME, 242 (1968) 1149-1160.

3. John D. Verhoeven, Fundamentals of Physical Metallurgy (New York, Ny: John Wiley & Sons, Inc., 1975) 339-342.

4. F. Haessner, ed., Recrystallization of Metallic Materials (Stuttgart: Dr. Riederer Verlag GmbH, 1978) 15.

5. P.N. Richards, "Low temperature recovery prior to recrystallization of Al-killed deep drawing steels", The Journal of the Australian Institute of Metals, 12 (1) (1967) 2-9.

6. W.C. Leslie et al., "Solution and Precipitation of Aluminium Nitride in Relation to the Structure of Low Carbon Steel", Transactions of the ASM, 46 (1954) 1470-1499.

7. J. de Boer, v.d. Geest, Hoogovens Group BV, 1970.

8. M. Kretschmer, "Experiences with the determination of the binding conditions of nitrogen in steels by hot extraction with hydrogen", Arch. Eisenhüttenwes. 46 (10) (1975) 649-654.

9. Jean D'Ans and Ellen Lax, Taschenbuch für Chemiker und Physiker (Berlin: Springer Verlag, 1949) 1113.

10. R.W.K. Honeycombe, Steels - Microstructure and Properties (London: Edward Arnold Ltd., 1981) 7.

11. F. Haessner, ed., Recrystallization of Metallic Materials (Stuttgart: Dr. Riederer Verlag GmbH, 1979) 140-142.

VERIFICATION OF A FINITE DIFFERENCE MODEL FOR THE DISSOLUTION OF Mg_2Si IN 6XXX ALUMINUM ALLOYS

L. A. Lalli and R. A. Dauer

Alcoa Laboratories
Alcoa Center, PA 15069

Abstract

The dissolution kinetics a series of 6XXX alloys have been quantified by differential scanning calorimetry and quantitative metallography. The as-cast microstructure is shown to be linearly related to the solvus temperature. The evolution of the microstructure is followed during an 8-hour heat-up to the final soak temperature. The segregation in both low and high solute alloys is shown to first increase and subsequently decrease during the heat-up. These measured kinetics are then compared to the predictions of a finite difference model that considers the interdiffusion coefficients of both magnesium and silicon in aluminum, but with no interaction terms. Local equilibrium as defined by the Al-Mg-Si solvus curve is assumed to exist at the interface of the second phase and the matrix. The effect of the temperature-time curve prior to achieving soak on the growth and dissolution of Mg_2Si is shown to be predicted reasonably well. It is concluded that the cell size distribution must be included in the model in order to improve the predictions.

Introduction

Prior to most industrial hot working processes cast ingots are homogenized in a "preheat" operation to dissolve any phases with low melting points that might affect adversely their workability, or require an extended solution heat treatment in the final product. Preheating furnaces are expensive and require considerable energy; consequently for their efficient utilization it is desirable to know the time-temperature kinetic relationship for achieving a single phase solid solution. This is particularly important for those alloys with a high solute content where the minimum temperature required for achieving complete solubility is very near the lowest eutectic melting temperature. The type and amount of second phases present in the as-cast structure will be determined by the chemical composition of the alloy and the solidification conditions; these also affect the primary and secondary dendrite arm spacings which determine the diffusion distances. In addition to the initial microstructure, the response of an alloy in a preheating furnace is also affected by the overall dimensions of the ingot as well as the characteristics of the furnace which determine the heat-up rate and thus the amount of dissolution prior to achieving the final soak temperature. It will be shown that the heat-up can be very important since first growth and then partial dissolution of second phases occurs before the metal reaches the intended soak temperature. In order to study the kinetics of dissolution in such a non-isothermal environment, a numerical finite difference technique will be utilized to solve the diffusion equations. Prior work (1) in the binary Al-Cu system demonstrated that reasonably accurate predictions could be expected for a single element diffusing in aluminum if the distribution of cell sizes in the as-cast microstructure was considered. The present work extends this model to two diffusing species, magnesium and silicon, in aluminum, and has a similar formulation to the work of Moren et al. (2). The evolution of the as-cast microstructure will be followed by both optical microscopy and differential scanning calorimetry (DSC).

Experimental Procedure

Material from the center of production size (406 to 610 mm minimum dimension) 6061 alloy direct-chilled semi-continuously cast ingots was used for the kinetic studies during preheating. In addition, in order to assist in developing a method for characterizing the initial microstructure as a function of composition and solidification conditions, unidirectionally solidified ingots were produced. While the intention was to use the latter for preheating studies also, this has not been done as yet. Table I summarizes the chemical composition of the various ingots. The data characterizing the as-cast microstructures is contained in Table II. Mean cell size determinations have been made by the line intercept method using five random lines at a magnification of 250X. Volume percent determinations have been made by counting the number of intersections of Mg_2Si with a 5 x 5 grid; 25 fields at 20X magnification were examined for each measurement. The differential scanning calorimetry data revealed the presence of either four or five reactions. Only the reaction with the highest temperature corresponding to the melting of the Al-Mg_2Si eutectic has been considered. Reproducibility of the DSC data was excellent, and only the average values of these data are given. An estimate has been made of the solvus temperatures by correcting for the amount of silicon that combines with iron and manganese and is thus not available for diffusion into the matrix. In the unidirectionally solidified ingots, the measurements of samples farthest from the chill corresponded most closely to the data from the production size ingots.

Table I. Chemical Composition

	Si	Fe	Cu	Mn	Mg	Cr	Ni	Zn	Ti
Unidirectional Ingots									
502425	.78	.20	.30	.00	.74	.18	.00	.01	.02
6	.49	.18	.31	.00	.70	.19	.00	.01	.02
7	.46	.19	.29	.00	1.01	.16	.00	.01	.02
8	.42	.18	.28	.00	1.24	.19	.00	.01	.02
9	.48	.19	.29	.00	1.50	.18	.00	.01	.02
Full-Scale Ingots									
502400	.42	.38	.28	.00	1.13	.18	.00	.01	.02
1	.42	.41	.27	.00	1.23	.19	.00	.01	.00
2	.63	.43	.30	.00	1.10	.18	.00	.01	.00
3	.65	.45	.31	.00	1.08	.19	.00	.01	.00
4	.71	.44	.34	.00	.96	.19	.00	.01	.01
502446	.69	.44	.32	.00	.94	.19	.00	.01	.00
500847	.72	.42	.39	.01	.93	.20	.00	.02	.04
849	.71	.40	.33	.02	.91	.19	.00	.01	.04
850	.67	.41	.33	.01	.88	.17	.00	.02	.03
859	.76	.40	.33	.00	.98	.17	.00	.01	.01
860	.69	.45	.40	.01	.83	.17	.00	.02	.07
486187	.65	.40	.30	.02	.89	.18	.00	.02	.03
229	.63	.33	.32	.00	.97	.19	.00	.02	.01
232	.65	.41	.31	.02	.89	.19	.00	.03	.03

Table II. Optical and DSC Data for As-Cast Structure

	Effective* Si	Solvus** (°K)	Cell Size 12 mm From Chill (mm)	Volume % at Distance From Chill		DSC (cal/gm) at Distance From Chill		
				200 mm	12 mm	200 mm	100 mm	12 mm
Unidirectional Ingots						Average of 2 Data Points		
502425	.75	818.2	.106	1.12	1.28	.8184	.8076	.3135
6	.46	795.4	.100	.32	2.24	.3413	.2025	.0398
7	.43	820.8	.089	.96	2.72	1.155	.8765	.2880
8	.39	833.5	.093	1.12	2.72	1.836	1.458	.7579
9	.45	855.3	.096	1.76	3.68	3.356	2.929	2.078
Full-Scale Ingots						Single Data Point		
502400	.36	822.4	-		-		1.6748	
1	.35	828.6	-		-		1.9772	
2	.56	838.4	-		-		2.2977	
3	.57	838.1	-		-		2.4113	
4	.64	832.6	-		-		2.0307	
446	.62	829.6	-		-		1.9393	
						Average of 3 Data Points		
500847	.65	830.8	.081		.59		2.180	
849	.64	828.5	.079		.84		1.830	
850	.60	823.2	.098		.61		1.710	
859	.69	837.8	.091		.81		2.148	
860	.61	819.4	.092		.42		1.212	
486187	.58	822.8	.131		.42		1.307	
229	.57	829.3	.120		.50		1.780	
232	.58	822.7	.121		.56		1.528	

*Effective Si = $Si - \frac{Fe + Mn}{6}$

**Solvus (°K) = 17013/[19.9 - 2 ln (wt.% Mg) - ln (wt.% Si)]

Samples from the production size ingots were heated in an air furnace that was programmed to follow a linear trajectory in terms of temperature versus the logarithm of time (3). Table III summarizes the coordinates used for the heat-up to the three soak temperatures. During the heat-up, samples were periodically removed and quenched. The results of DSC and optical microscopy for these samples are given in Tables IV and V.

Model Description

Since most aspects of the present model are similar to others in the literature, only a brief description will be given here. Rectangular coordinates are used throughout this work. The interdiffusion of the two elements, Mg and Si, in aluminum is modeled using Fick's second law of diffusion with temperature dependent diffusivities (4); thus, it is assumed that concentration gradients provide the driving force toward achieving equilibrium. Additionally, the matrix concentration gradient of one element is assumed not to affect the diffusion flux of the other; i.e., interaction terms are ignored. However, the concentrations of the two elements are directly coupled at the interface between the particle and the matrix; here the Al-Mg-Si solvus equation given at the bottom of Table II is used as a boundary condition. Thus, the local matrix concentrations at the interface are assumed to be in equilibrium (or fully saturated) at the instantaneous temperature. The Mg_2Si is modeled as a single particle with a constant stoichiometry (63.4 wt% Mg, 36.6 wt% Si). The velocity of the interface between this particle and the matrix is determined by solute balances for the two elements; as shown in Figure 1, in order for the interface to remain distinct, both solute balances must result in the same interface velocity. Thus, an iterative procedure is used at each time step to solve for the interface velocity and the matrix interface concentrations, C_{s1} and C_{s2}. The resulting interface velocity is used then to determine both the subsequent size of the particle and the updated coordinates of the moving finite difference grid. This moving grid reference frame (5) is fixed to the plane of symmetry between two particles on one end and, at the other end, is fixed to the particle-matrix interface; at intermediate positions, it moves with a velocity that is proportional to its distance from both ends. Thus, the grid accounts for the increase of matrix material as the particle dissolves; for small volume fractions, however, this effect is small. Mathematically, convection terms must be added to the diffusion equation in such a moving reference frame. This is similar to the solute balance that occurs at the interface. The initial conditions of volume fraction Mg_2Si are determined by the optical data in Table II, with the matrix concentration assumed to be uniform and at a level as determined by the amount of Mg_2Si and the alloy's composition.

Figure 2 gives some insight into the factors affecting the growth and dissolution kinetics as predicted by this model. The interfacial matrix concentrations for Mg and Si are shown to increase with time for the 8-hour heat-up to 554°C; these are determined by the solvus equation but are affected by the solute fluxes. The matrix concentrations at the zero flux boundary are also shown. The difference between the two may be thought of as an effective driving force for the growth and dissolution of the particle, the rate of which will also depend on the diffusivities which vary with temperature.

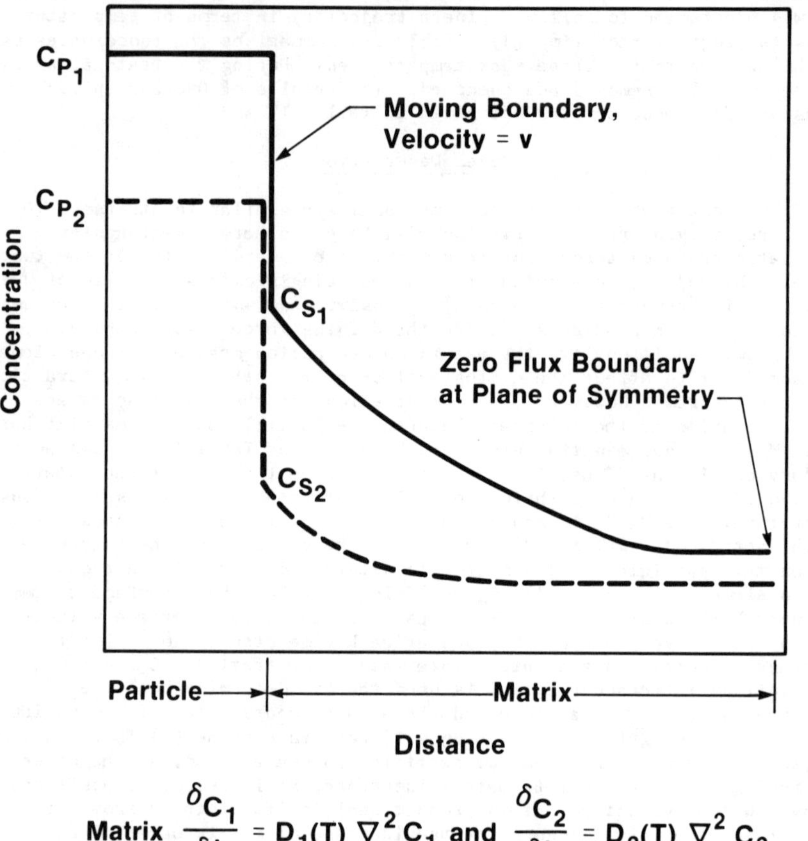

Matrix $\frac{\delta C_1}{\delta t} = D_1(T)\, \nabla^2 C_1$ and $\frac{\delta C_2}{\delta t} = D_2(T)\, \nabla^2 C_2$

Particle C_{P_1}, C_{P_2} = Constant

Particle - Matrix Interface

$$(C_{P_1} - C_{S_1})v = D_1 \delta C_1/\delta x$$
$$(C_{P_2} - C_{S_2})v = D_2 \delta C_2/\delta x$$
$$f(C_{S_1}, C_{S_2}, T) = 0 \text{ (Solvus Equation)}$$

Figure 1
Description of Model

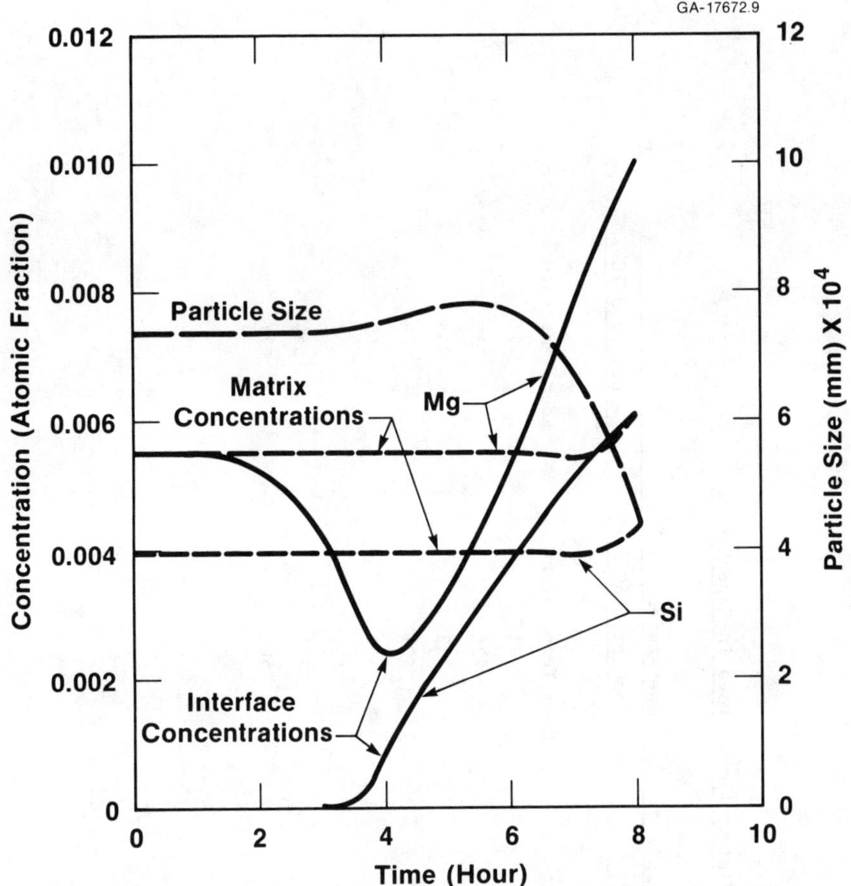

Figure 2

Predicted Interface and Matrix Concentrations for S-500859 During 8 Hour Heat-Up to 554°C

Table III. Heat-Up Curves Used in Experiments

Time (hr)	Temperatures (C)		
	Soak at 554C	Soak at 566C	Soak at 577C
0	Room Temperature	Room Temperature	Room Temperature
1	71	71	71
2	232	236	239
3	328	332	338
4	396	401	408
5	445	454	462
6	488	497	507
7	528	534	544
8	554	566	577

Discussion of Results

As a method of comparing the variations in the as-cast microstructures with composition and solidification conditions, the data in Table II has been related to the solvus temperature as shown in Figures 3 and 4. It is apparent that the amount of Mg_2Si increases with composition in a linear fashion with the solvus temperature. There is not necessarily any direct physical significance to this correlation since the volume fraction of eutectic remaining after solidification, if equilibrium were maintained at the solid-liquid interface, is expected to be determined by the liquidus and solidus surfaces and the tie lines between the two, as well as any diffusion during or subsequent to solidification. It is noted that the cooling rates during solidification in the unidirectionally solidified ingots were measured to be approximately 5, 2, and 0.5 C/s at 12, 100, and 200 mm from the chill, respectively. The volume percent data in Figure 3 indicates different relationships to the solvus temperature for the unidirectional versus full scale ingots; it should be noted that the data for the two ingot types was obtained on different occasions. Additionally, since the DSC data does not show this difference, and since the cooling rates for the two ingot types are believed to be similar, the DSC data is assumed to be the more accurate. Figure 5 shows that a good correlation between the optical and DSC data is possible for the production scale ingots in both the as-cast and preheated conditions. This correlation gives some confidence that DSC may be used to follow the microstructural changes during preheating. It may also be seen in Figure 5 that (1) both the optical and DSC data decrease significantly with only 0.1 hour at 554°C (following an 8 hour heat-up), and (2) 2 hours at 554°C result in similar undissolved eutectic as 0.1 hour at 577°C. A correlation between the optical and DSC data is also given in Figure 5 for the production scale ingots. As a check on its accuracy, this correlation was then used to predict the average matrix concentration for the unidirectional ingots. This calculation is compared to microprobe data for low and high solute alloys in Figure 6. Since the comparison is reasonably good, this further supports the use of DSC and the volume percent correlation to prescribe the initial conditions for the model. The microprobe data shown in Figure 6 also reveals the expected coring across the dendrite cells; this was not included in the calculations presented here since it has a small effect on the total dissolution time.

In order to demonstrate the use of DSC for studying the dissolution kinetics of samples prior to achieving the desired soak temperature, data from three ingots (low, medium, and high solute contents) in Table IV is plotted in Figure 7, where it is seen that growth of Mg_2Si occurs during the heat-up before dissolution commences. Complete dissolution also occurs sooner for the low solute alloys, at times before ever achieving the prescribed soak temperature. The faster homogenization of low solute alloys is a result of less as-cast eutectic and a lower matrix concentration in the final stages of dissolution resulting in higher solute fluxes. Figure 8 is a plot of all of the data contained in Table IV but has been normalized by dividing the data by the as-cast DSC measurements. The remaining scatter in the data indicates that this normalization cannot account for all of the differences between alloys, but this will be shown to be a useful method of comparing the DSC data to the predictions of the model.

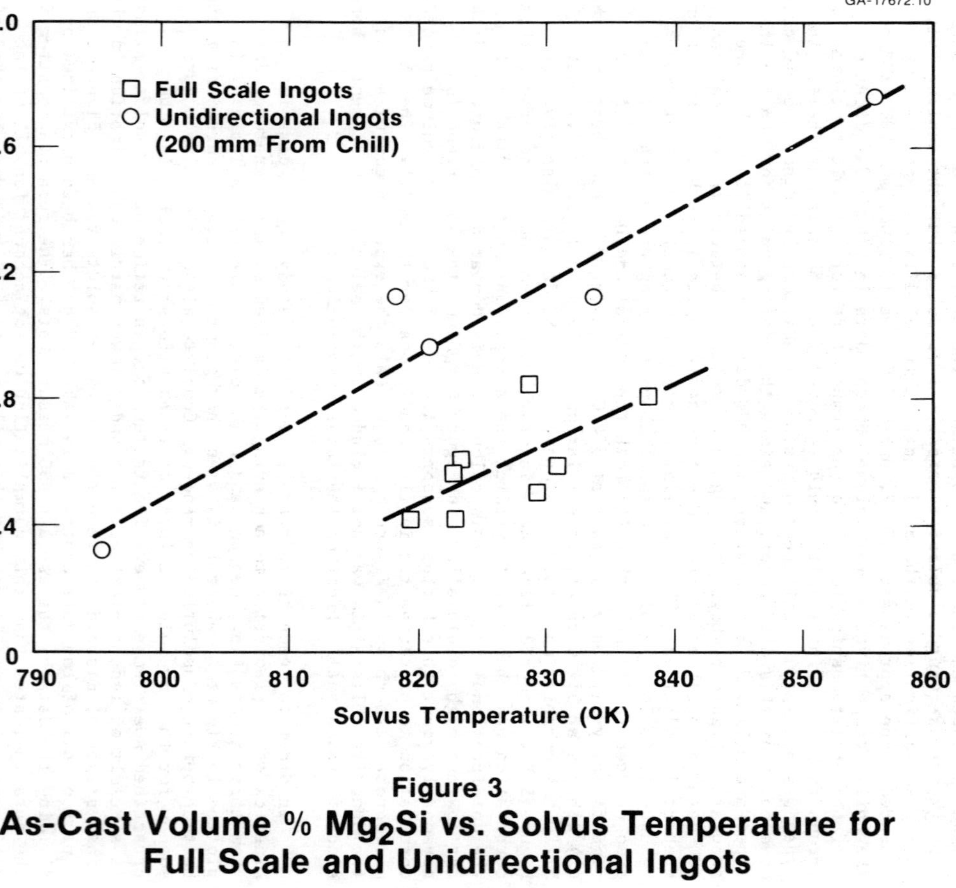

Figure 3

As-Cast Volume % Mg_2Si vs. Solvus Temperature for Full Scale and Unidirectional Ingots

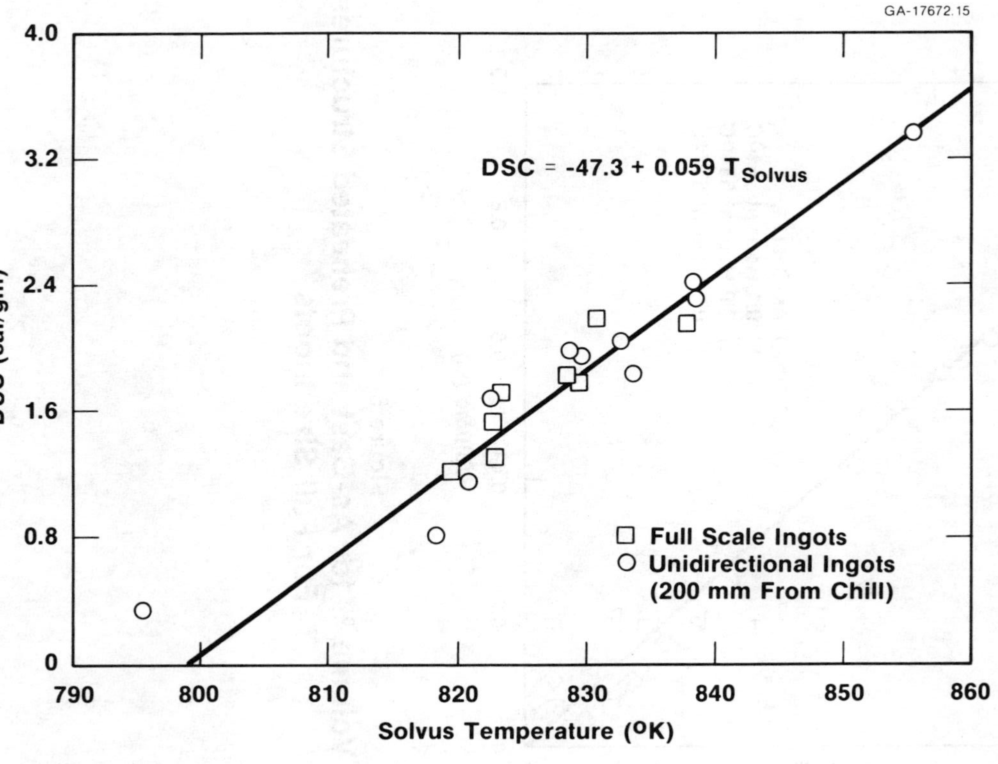

Figure 4

As-Cast DSC Measurements vs. Solvus Temperature for Full Scale and Unidirectional Ingots

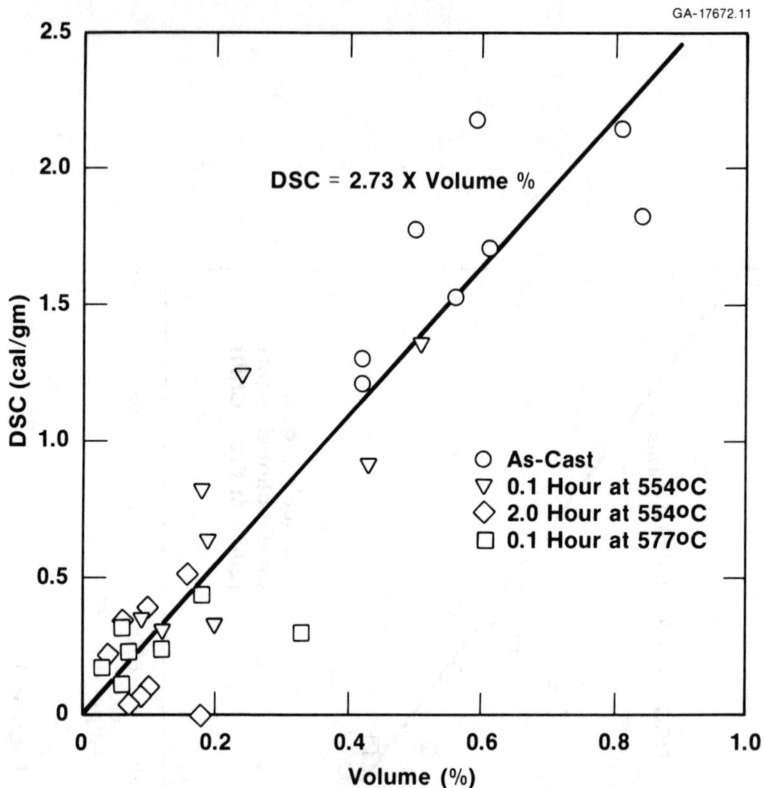

Figure 5

DSC vs. Volume % for As-Cast and Preheated Structures For Full Size Ingots

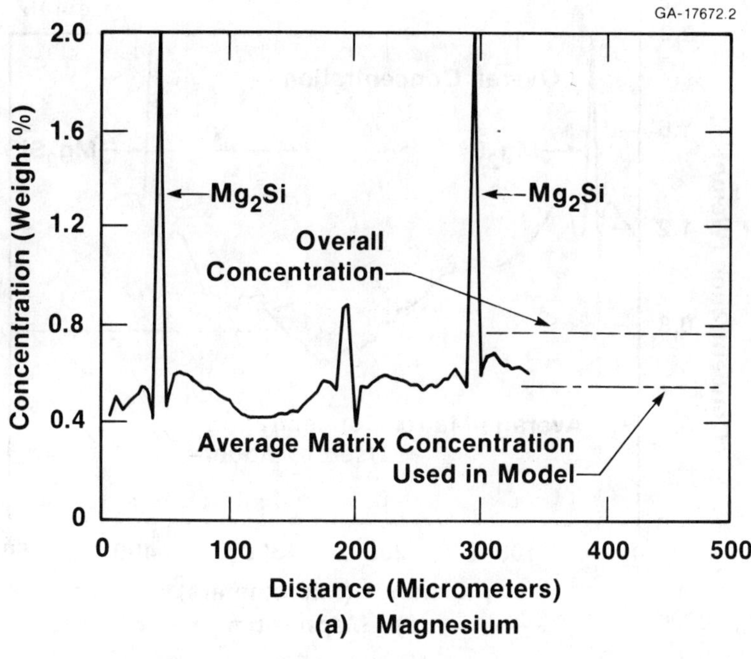

(a) Magnesium

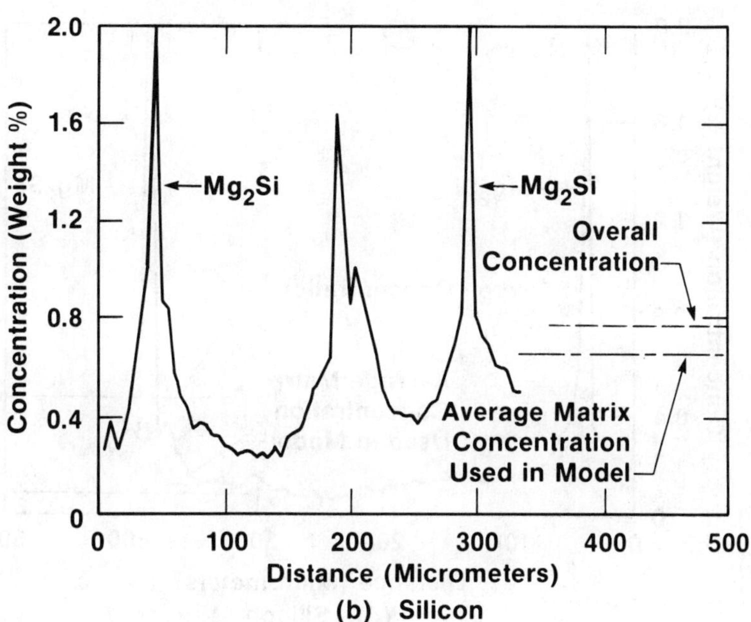

(b) Silicon

Figure 6 a & b

Concentrations in S-502425

(200 mm From Chill)

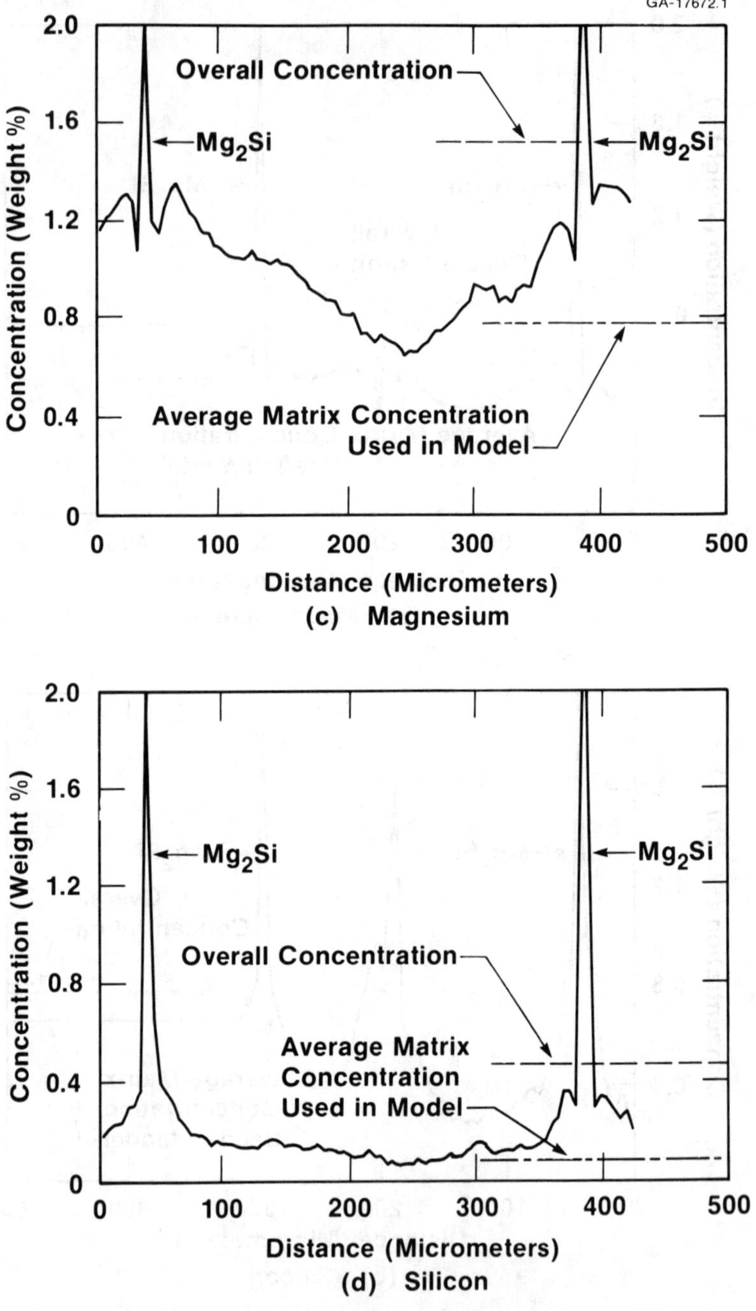

Figure 6 c & d

Concentrations in Ingot S-502429

(200 mm From Chill)

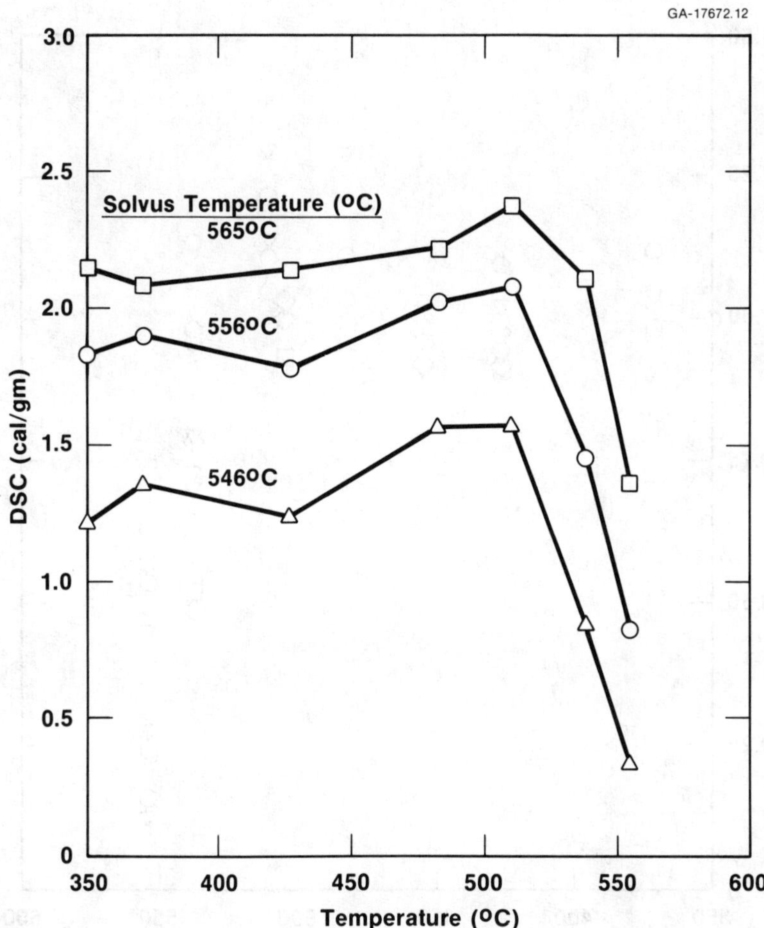

Figure 7
DSC During 8 Hour Heat-Up to 554°C for Low,Medium, And High Solute Alloys

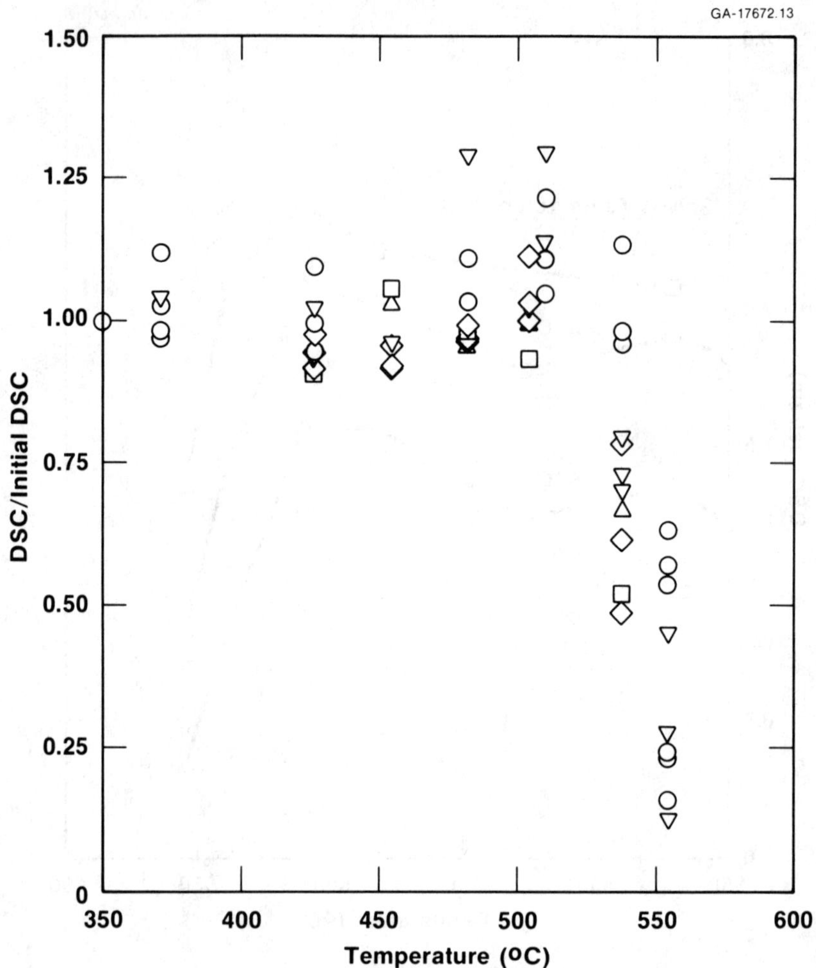

Figure 8
Normalized DSC During 8 Hour Heat-Up to 554°C for Full Scale Ingots

Table IV. Optical and DSC Data for Ingots Following an 8-Hour Heat-Up to 554°C

Data From Samples Quenched on Achieving the Indicated Temperatures

Full-Scale Ingots	371C (cal/gm)	427C (cal/gm)	454C (cal/gm)	482C (cal/gm)	504C (cal/gm)	538C (cal/gm)	554C (cal/gm)	0.1 Hour at 554C (cal/gm)	0.1 Hour at 554C Volume %	2 Hours at 554C (cal/gm)	2 Hours at 554C Volume %
502400		1.6363	1.5417	1.6134	1.7290	1.0301					
1		1.8666	1.8148	1.9150	1.9759	.9635					
2		2.1612	2.2105	2.5556	2.3570	1.6734					
3		2.2057	2.3035	2.3895	2.6827	1.8880					
4		1.9570	2.0913	1.9399	2.0243	1.3527					
446		1.7617	2.0491	1.9004	1.8098	1.0091					
					(510C)						
500847	2.2367	2.3861		2.4187	2.6515	2.0915	1.2072	1.2440	.24	.3429	.06
849	1.8988	1.7806		2.0247	2.0805	1.4540	.7669	.8258	.18	.2237	.04
850	1.6831	1.6187		1.8784	1.7905	1.9394		.9178	.43	.3951	.10
859	2.0817	2.1411		2.2200	2.3811	2.1093		1.3576	.51	.5156	.16
860	1.3554	1.2382		1.5636	1.5695	.8448		.3340	.20	.0000	.18
486187								.3520	.09	.0417	.07
229								.6397	.19	.1002	.10
232								.3086	.12	.0714	.09

Fully Homogenized in 4 Hours

The model's predictions for the heat-up and soak at 554°C is shown in Figure 9 as compared to the normalized DSC data, where data greater than 1.0 indicates the predicted growth prior to dissolution. A reasonably good fit results when the cell size is adjusted upward by a factor of 2. Physically, one reason for this could be that there is actually a range of cell sizes that control the diffusion distance. As can be seen in Figure 10 for a binary Al-Cu alloy, the cell size distribution significantly affects the kinetics. Using the average cell size underpredicts the dissolution kinetics near the end of homogenization where the largest diffusion distances are controlling. In the present work, it is assumed that a single cell size, larger than the average by a fixed percentage, may be used to compare the microstructure's response to various temperature-time conditions. Figure 11 demonstrates the sensitivity of the model's predictions to changes in the assumed average diffusion distance. It should be noted that the model predictions are essentially identical if the diffusivities for both elements are reduced by a factor of 4, instead of changing the diffusion distance. The diffusivities used are shown below (4) and are believed to be accurate for this alloy:

$$D(T) = D_0 \exp(-Q/RT),$$

where for Mg, $D_0 = 0.063\ cm^2/s$ and $Q = 27$ kcal/mol,
and for Si, $D_0 = 18.0\ cm^2/s$ and $Q = 36$ kcal/mol.

The model's predictions for the data contained in Table V for the soak temperatures of 566°C and 577°C is shown in Figure 12, in addition to the 554°C data. Although there is limited data at the higher temperatures, the comparison appears equally good. Figure 13 contains predictions for ingots where the cell size was not measured and therefore was assumed. In addition, the calculated solvus temperatures for most of these ingots was in excess of the 554°C soak temperature. Consequently, the predicted particle sizes will not approach zero. The fact that the measured DSC data does approach zero indicates that either the solvus equation or the estimate of effective silicon that is used in this equation may be slightly in error. While the predictions do correlate with the data, the fit is not nearly as good in an absolute sense. It is, therefore, critical to accurately represent the diffusion distances and the solvus equation for accurate predictions.

Figure 14 shows how the model may be utilized for examining preheating practices where the effect of solute level on the time to achieve complete dissolution is shown. The heat-up curve is predicted from a simple exponential heat transfer calculation, but any curve including experimental data may be utilized. The effect of solute level on the initial as-cast particle size and on the subsequent kinetics can be seen. For the highest solute level, complete dissolution is never achieved since the final soak temperature is not above the solvus temperature for the alloy.

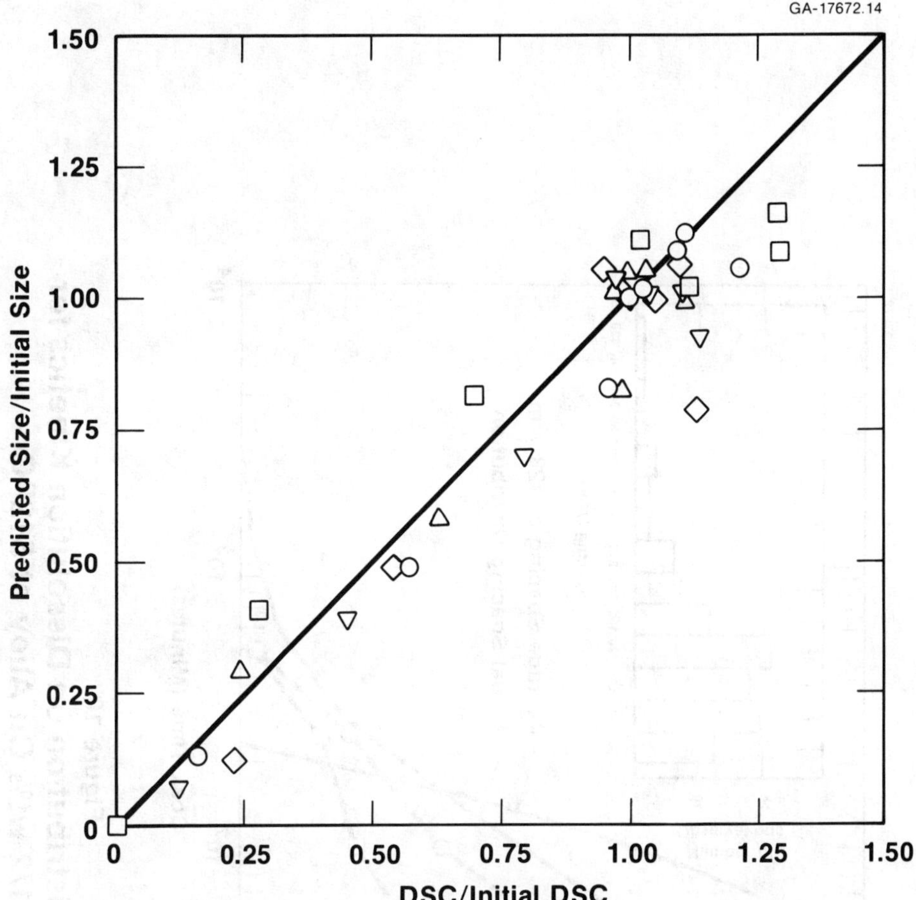

Figure 9
Model Predictions vs. DSC
During 8 Hour Heat-Up to 554°C
Samples S-500847-860, Diffusion Distance = 2.0 X Cell Size

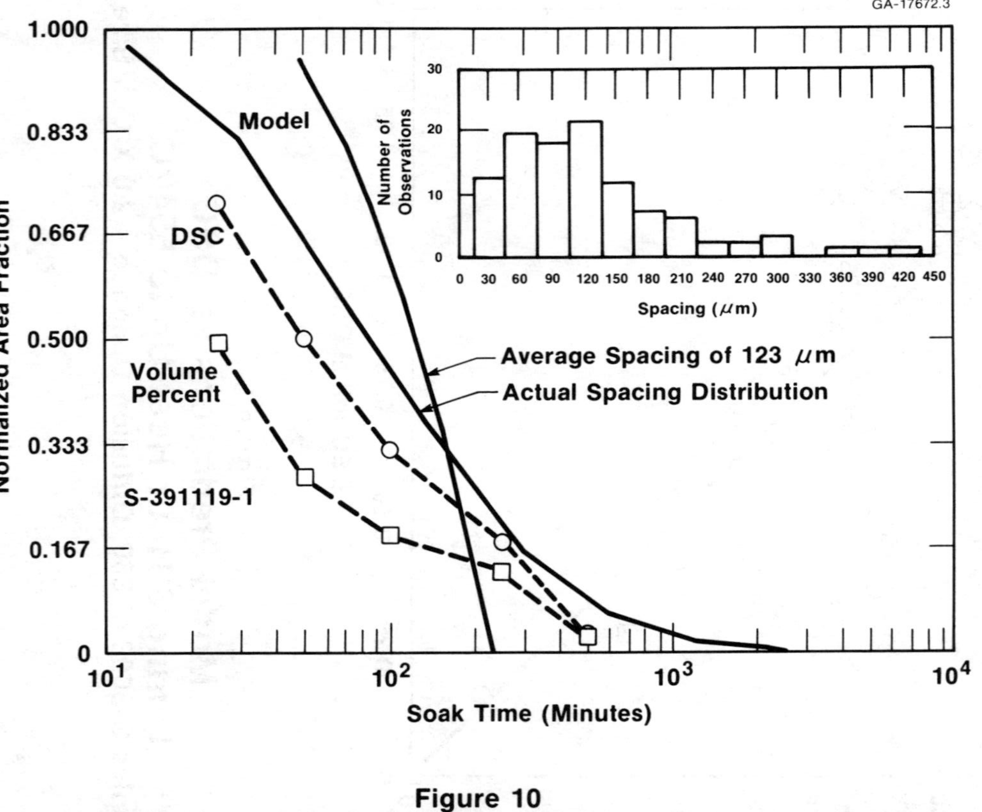

Figure 10

Effect of Cell Size Distribution on Dissolution Kinetics for An Al - 3 1/2 wt% Cu Alloy at 512°C

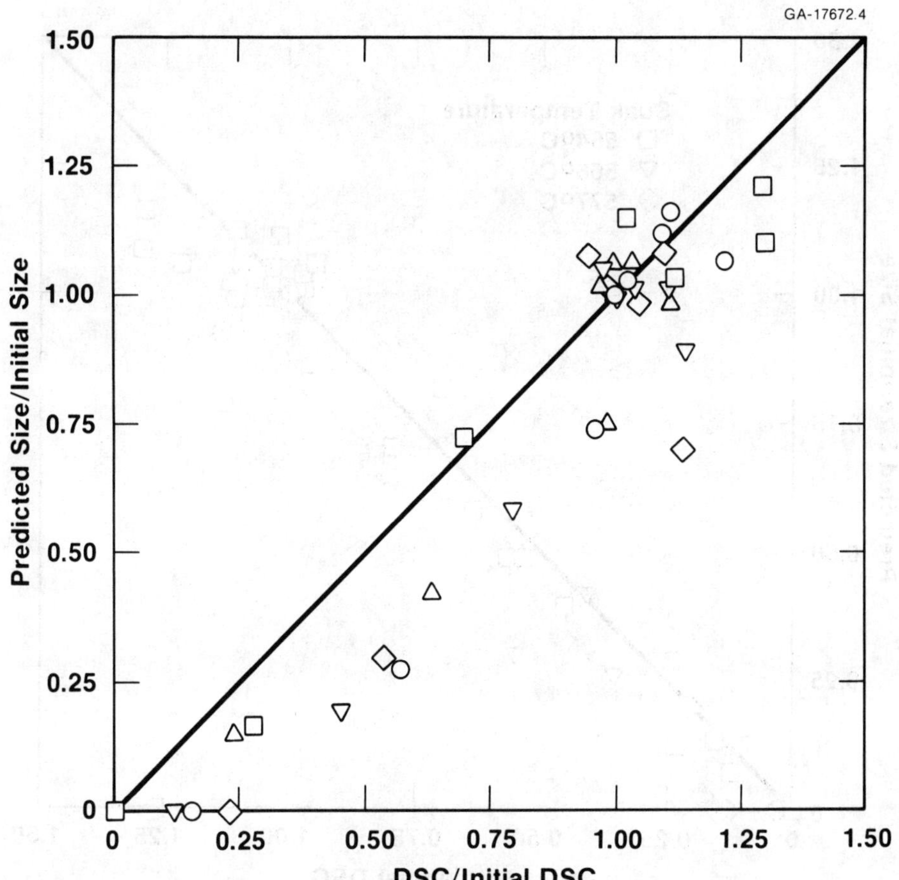

Figure 11

Model Predictions vs. DSC for 8 Hour Heat-Up to 554°C

Samples S-500847-860, Diffusion Distance = 1.5 X Cell Size

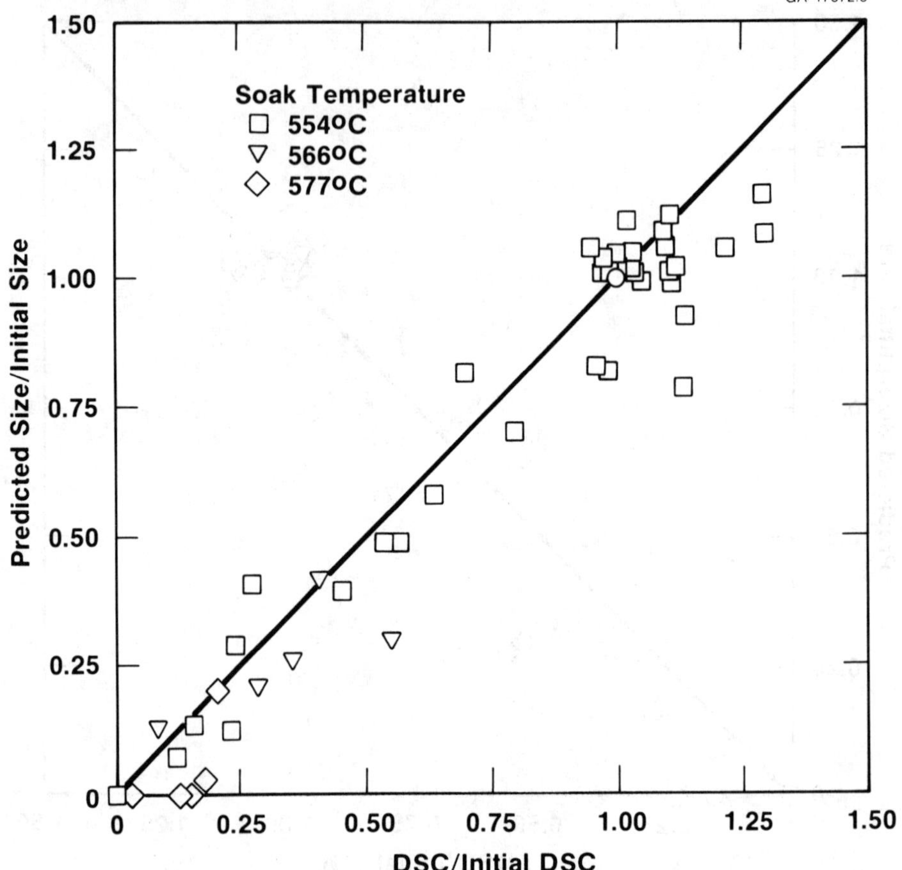

Figure 12

Model Predictions vs. DSC During 8 Hour Heat-Up to 554°C, 566°C, 577°C

Samples S-500847-860

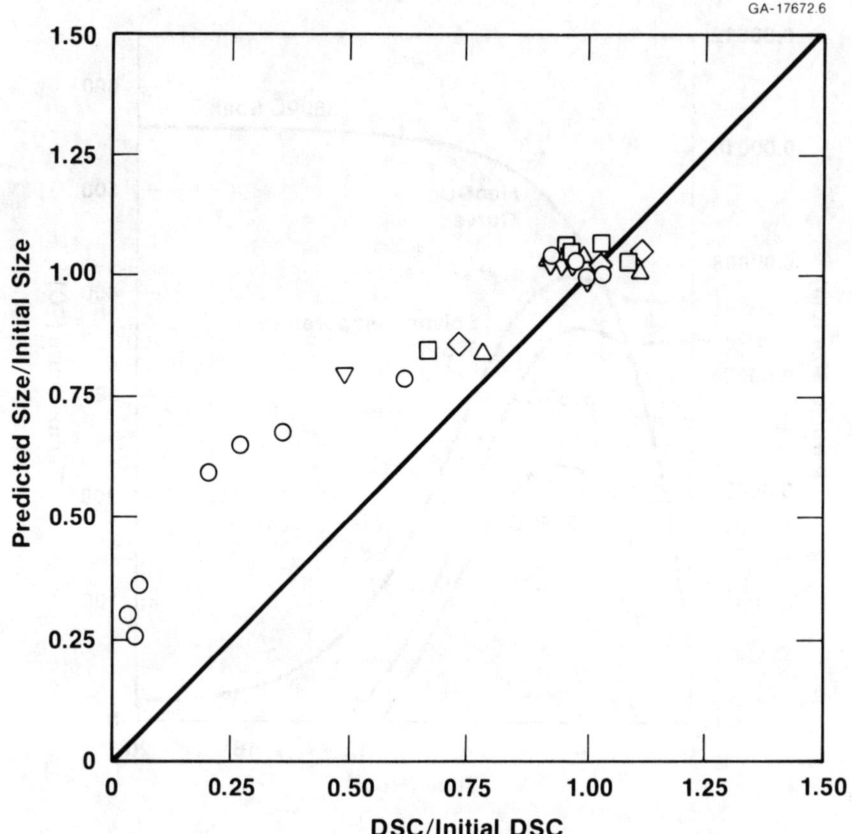

Figure 13

Model Predictions vs. DSC During 8 Hour Heat-Up to 554°C for Samples Where Cell Size Not Known

Samples S-502400-404,446, Diffusion Distance = 2.0 X Cell Size

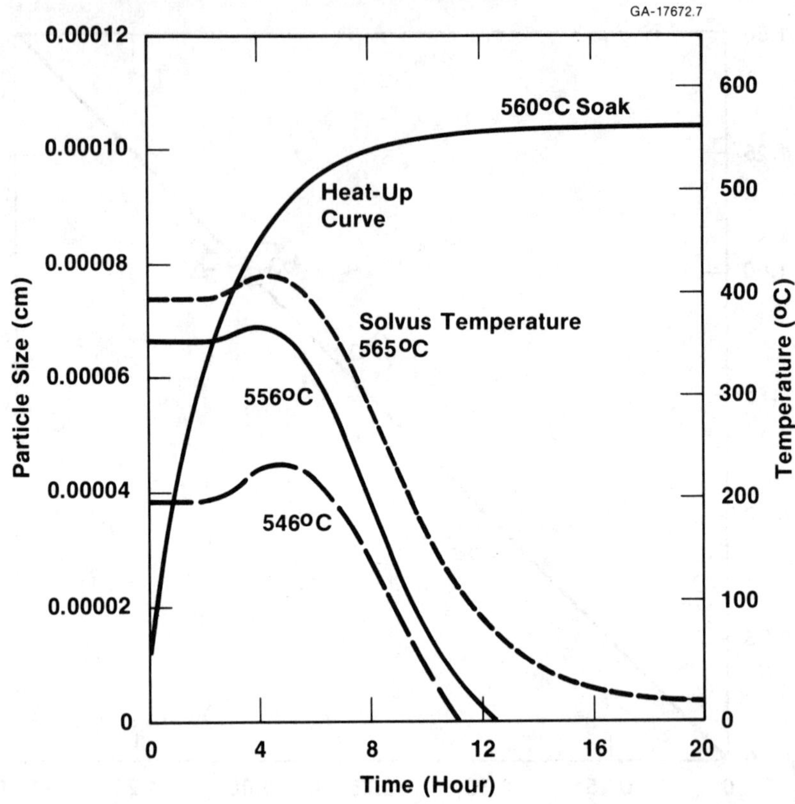

Figure 14

Predicted Dissolution Kinetics for Low, Medium and High Solute Alloys for Hypothetical Heat-Up Curve

Table V. Optical and DSC Data for Ingots Following an 8-Hour Heat-Up to Indicated Temperatures

	566C	0.1 Hour at 566C	577C	0.1 Hour at 577C	
	(cal/gm)	(cal/gm)	(cal/gm)	(cal/gm)	Volume %
500847	.7361	.7736	.3485	.3199	.06
849	.5467	.5236	.1760	.2402	.12
850		.9404		.3012	.33
859		.8770		.4398	.18
860		.1034		.0311	–
486187		.1105		.1113	.06
229		.2402		.2339	.07
232		.1221		.1721	.03
	Fully Homogenized in 2 Hours			Fully Homogenized in 1.6 Hours	

Conclusions

1. A diffusion model that considers two elements diffusing in an aluminum matrix has been shown to give reasonable, first order predictions for both the initial growth and subsequent dissolution of Mg_2Si for a range of solute levels under non-isothermal conditions.

2. Both the as-cast and partially homogenized microstructures have been quantified through optical and DSC techniques, and these two techniques were correlated to one another.

3. A correlation was developed relating the as-cast microstructure to the solvus temperature for Al-Mg-Si alloys.

4. Particle growth has been shown to occur for both low and high solute alloys during heat-up to soak temperature.

5. A more accurate characterization of the as-cast microstructure including the distribution of dendrite arm spacings is necessary to improve the predictions of the model.

Acknowledgements

The writers would like to thank D. A. Granger for producing the unidirectional ingots, H. C. Stumpf for the thermodynamic data, and R. B. Cunningham for the preheating data on the production size ingots.

References

1. L. A. Lalli, Internal Report, 12-79-FR205, 1979 November.

2. A. Moren, E. Randich, and J. I. Goldstein, Proceedings of International Conference in Nuclear Metallurgy, 20, Part 1, (1976).

3. R. B. Cunningham, Internal Report, HH86, 1980 July.

4. W. G. Fricke, Scripta Met., 6, (1972), pp. 1139-1144.

5. D. Murray and F. Landis, Trans. ASME, 81, (1959), p. 106.

COMPUTER SIMULATION OF QUENCHING

URANIUM-0.75 WEIGHT PER CENT TITANIUM ALLOY

G. M. Ludtka, G. H. Llewellyn, G. A. Aramayo, M. Siman-Tov, K. W. Childs

Martin Marietta Energy Systems, Inc.,
Oak Ridge Y-12 Plant
P.O. Box Y
Oak Ridge, Tennessee 37831

Abstract

A "QUENCH SIMULATOR" has been developed which uses finite difference heat transfer and finite element stress analysis techniques to predict the behavior of a metal during quenching. The actual nonlinear temperature- and microstructure-dependent physical, thermophysical, and mechanical properties are incorporated as input into the computer model as well as the continuous cooling transformation (CCT) behavior and heats of transformation of the alloy. The final output provides the transient temperature distribution, details the final residual stress profile, predicts and shows where distortion occurs, and maps out the microstructure distribution throughout the entire sample. These data are available in tabulated form, contour plots, or color-coded graphics.

This analysis has been demonstrated on simple shapes for unalloyed uranium and the uranium-0.75 weight per cent titanium alloy which undergoes a martensite transformation and is quench-rate sensitive. The results of this study will be discussed in detail in addition to other applications of this analytic approach which is generic in nature.

Introduction

Currently in industry, the determination of the effect of a heat treatment on an alloy is purely experimental. In particular, the influence of the quenching treatment on the microstructure, properties, and distortion of a part is obtained by a trial and error approach, i.e., a prototype part is manufactured, quenched, and subsequently evaluated to see if the component design and processing sequences produce the required effects in the quenched part. Frequently, several costly iterations are required before the final design configuration and processing procedures are finalized. The need for a predictive tool which would evaluate the influence of design, material choice, and processing history is clearly evident. The goal of this program was to establish such a tool which could be incorporated into an overall CAD/CAM approach to facilitate optimum design with the lowest cost and shortest lead time from conception to practice. The minimum output of this computer simulation was chosen to include predictions of:

1. transient temperature distributions,
2. residual stress profiles,
3. space and time dependent microstructure distributions,and
4. shape distortion.

Unalloyed uranium was chosen as an initial candidate to model since U undergoes two allotrophic phase transformations (alpha to beta, beta to gamma) whose heats of transformation would have to be incorporated into the overall heat transfer calculations. These data as well as other physical and mechanical properties were available in the open literature as a function of temperature. The more complex uranium-0.75 weight per cent titanium alloy was subsequently chosen for analysis since this alloy exhibits an additional martensite reaction which can form prior to time dependent phase decomposition upon cooling as summarized by a continuous cooling transformation diagram. Also, the U-0.75Ti alloy has been documented to develop centerline porosity under certain quench rate and geometry-related conditions. This information provided some background information of situations to model to determine whether the computer simulation would predict large residual stresses in those regions where void formation had been documented.

The Quench Simulator

Basic Software

The Quench Simulator is based on two commercially available software packages. The heat transfer calculations are accomplished using the HEATING6[1] finite difference three-dimensional heat transfer program, which is in the public domain. The distortions and residual stresses are obtained with the proprietary (licensing is available) finite element three-dimensional ADINA[2] program. These programs are very general and required the development of additional subroutines to facilitate:

1. simulation of the exothermic microstructure transformations whose transformation temperatures, and heats of transformation are complex functions of time and temperature;

2. simulation of nonlinear thermal boundary conditions encountered in the boiling process;
3. simulation of temperature and phase dependent physical, thermophysical, and mechanical properties,and,
4. simulation of moving boundary conditions to facilitate modeling of controlled immersion rate quench processes.

An overall view of the scope of this study is shown in Figure 1. This flow diagram shows the interrelationship of the experimental portion of this endeavor with the analytic modeling work. As can be seen for a given choice of materials, the physical and mechanical properties are critical input into the software programs. These data had to be determined for the U-0.75Ti alloy and required extensive testing since elevated temperature data were not known. In addition, the heat transfer calculations require inputting the appropriate surface heat transfer coefficient for the method of quenching. The data available in the literature for heat transfer data in water were only for equilibrium conditions, i.e., the test samples were kept at temperature and the data calculated. Immersion quenching in reality is a transient condition and so heat transfer data measured during the quench had to be experimentally determined. This work is almost complete. In the interim, a general boiling curve, shown in Figure 2, has been developed based on the most applicable steady state correlations available in the open literature. The plot in Figure 2 was used to provide the heat transfer boundary conditions for the water immersion quenching simulated in this study.

Materials Simulated

Two materials were chosen to be modeled. They are unalloyed uranium and the uranium-0.75 weight per cent titanium alloy. The microstructures of these will be briefly discussed to provide background information into the capability of the software to handle various microstructures.

The unalloyed uranium exhibits two allotrophic phase transformations. At room temperature and up to 667°C, an orthorhombic crystal structure is the equilibrium phase which is designated as the alpha phase. Between 667°and 775°C the beta phase exists which has a complex crystal structure of the tetragonal lattice type. The body centered cubic gamma phase forms above 775°C.

Upon dilute alloying with titanium, an additional equilibrium phase can be formed which is designated delta as shown in Figure 3. The delta phase is an intermetallic compound with composition U2Ti. For the U-0.75Ti alloy, the delta phase is stable as high as 723°C under equilibrium conditions. The microstructures of interest for this study are those predicted by a continuous cooling transformation diagram for quench rates faster than 1°C per second as presented in Figure 4. At the fastest quenching rates, a alpha prime martensite transformation occurs. The various types of microstructures that can evolve as a function of quench rate are shown in Figure 5. At the fastest rates, an acicular martensite is developed. As the quench rate is reduced, the microstructure evolves into the equilibrium alpha plus delta microstructure. This alloy exhibits discontinuous precipitation at grain boundaries at the slower rates. The effect of isothermal aging on the phase transformations in gamma solution heat treated and water quenched U-0.75Ti is summarized in Figure 6.

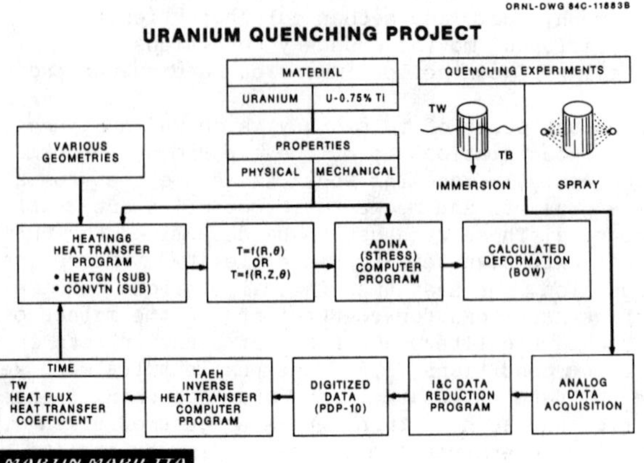

Figure 1. A summary of the interrelationship of the analytical and experimental efforts for the uranium alloy quenching simulation program.

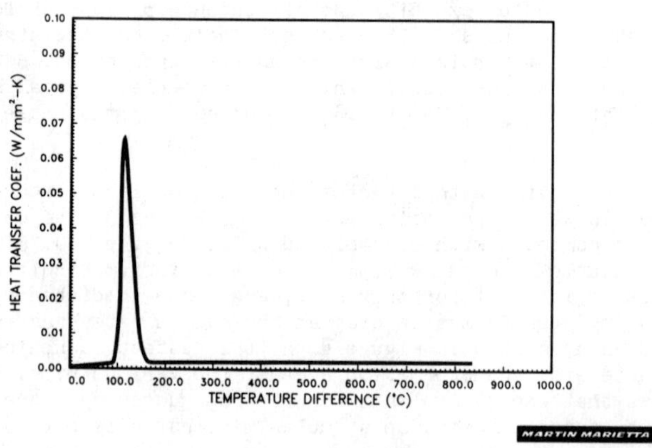

Figure 2. Linear plot of the surface heat transfer coefficient as a function of the wall temperature difference (Tw-Tb) where Tw is the wall surface temperature and Tb is the bulk temperature of the quenching bath. This data represents a compilation of steady state data available from the literature.

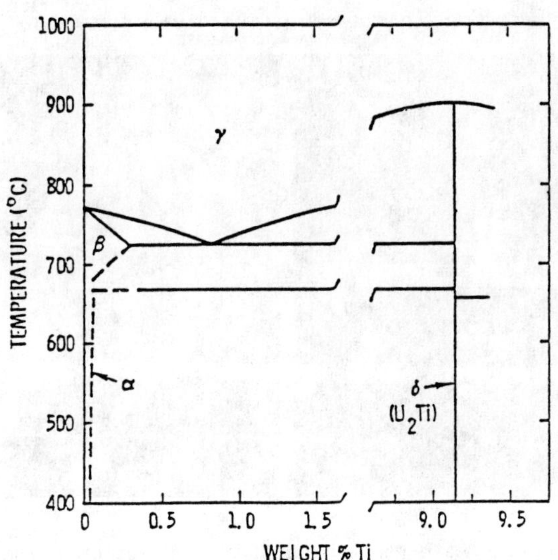

Figure 3. Uranium-Titanium equilibrium phase diagram.

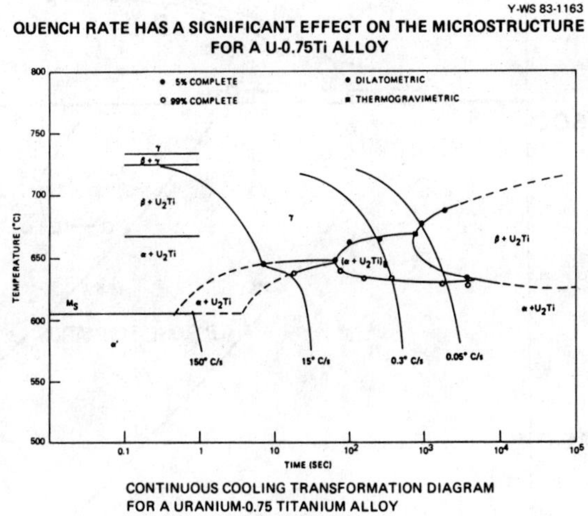

Figure 4. The Continuous Cooling Transformation diagram for a uranium-0.75 wt% titanium alloy.

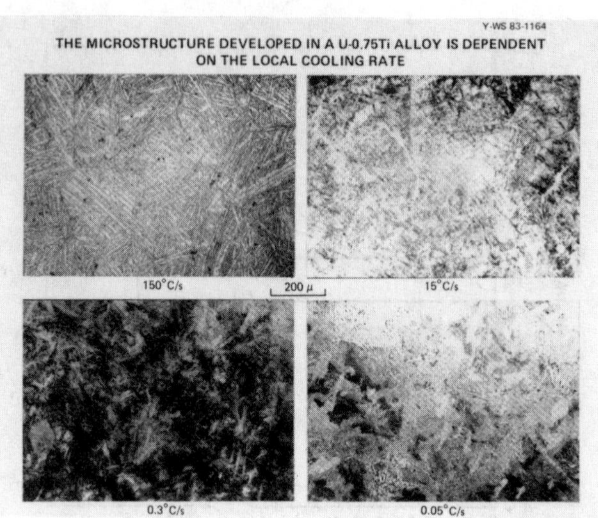

Figure 5. Cooling rate has a significant effect on the microstructure that evolves for the uranium-0.75% titanium alloy as summarized by these photomicrographs.

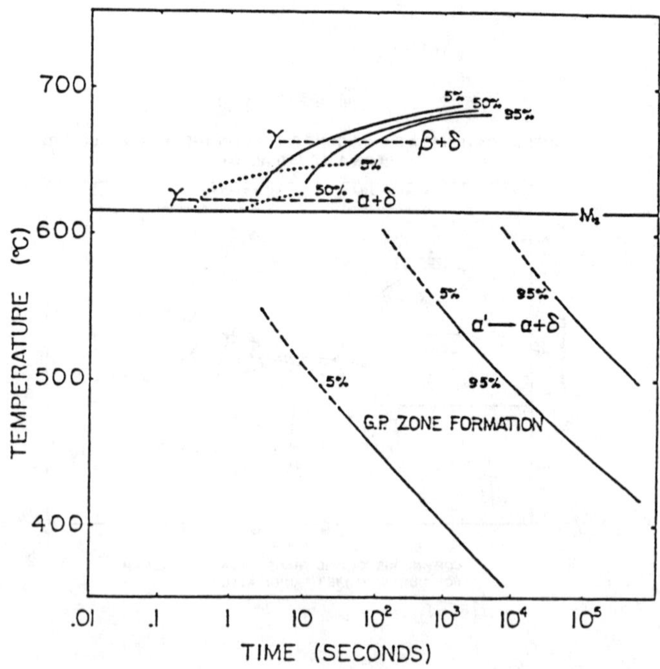

Figure 6. The isothermal transformation diagram for the uranium-0.75% titanium alloy.

Preliminary Considerations

Several major considerations had to be addressed to successfully model the quenching process. The first to be overcome was the necessity to accurately simulate the complicated time and temperature dependent phase transformations that occur during a quench as summarized in the CCT curve. This information is essential to the model since the heat of transformation associated with the phase changes that occur can have a significant impact on the cooling rate and consequently the amount of martensite formed for the U-0.75Ti case. In the modeling of the quenching using the HEATING6 heat transfer computer code, these heat of transformations are handled in a user supplied subroutine HEATGN. HEATGN allows the modeling of a heat generation function that is dependent on both time and temperature. In the HEATGN subroutine written for this case the fraction of each phase present is determined from the CCT diagram shown as Figure 4. For a given node in the model, the amount of each phase present can be read from the map as a function of the node temperature, the time since the node passed the critical temperature, and the node's past history. Based on this phase information the amount of heat of transformation which is released during the current time step is determined and entered into the problem as a heat generation. Subroutine PHSMAP keeps up with the phase distribution at any time and prints phase maps as requested. The relative magnitude of the contribution of the heat of transformation to the overall heat that must be removed by the quench process is shown in Figure 7 for unalloyed uranium. Approximately 30% of the heat extracted upon cooling can be directly related to the heats of phase transformation alone.

Another prime consideration that had to be accounted for was the rate at which the quenching material was transferring its heat content into the surrounding quenching medium. This information is normally represented by specifying the heat transfer coefficient. Numerous factors affect the value obtained for the heat transfer coefficient due to the complexity of the surface phenomenon occurring at the quench water-specimen interface. Ideally, experimentally determined data should be used as input to define the boundary conditions during the quenching process. However, these data were not available for uranium or its alloys. The literature was reviewed and a preliminary "universal" relationship was developed from the existing literature and is shown in Figure 2. This surface heat transfer curve was used as input for the parametric work to be discussed subsequently. Concurrently, an experimental program was initiated to obtain the surface heat transfer coefficient data for unalloyed uranium and the U-0.75Ti alloy. These surface heat transfer coefficient data is obtained by accurately measuring the transient temperature distribution evolved at several locations in a quenching sample and back calculating the surface heat flux, surface temperature, and surface heat transfer coefficient through an inverse heat transfer code called TAEH using a solution technique described by Beck.[3] The experimental portion of this work has been completed and the data reduction is underway.

A big constraint at the initiation of this endeavor was the paucity of data available for the physical, thermophysical, and mechanical properties for the U-0.75Ti alloy as a function of temperature and microstructure. An extensive experimental program was undertaken to obtain the required data for both the thermal and stress calculations. Figure 8 summarizes the data

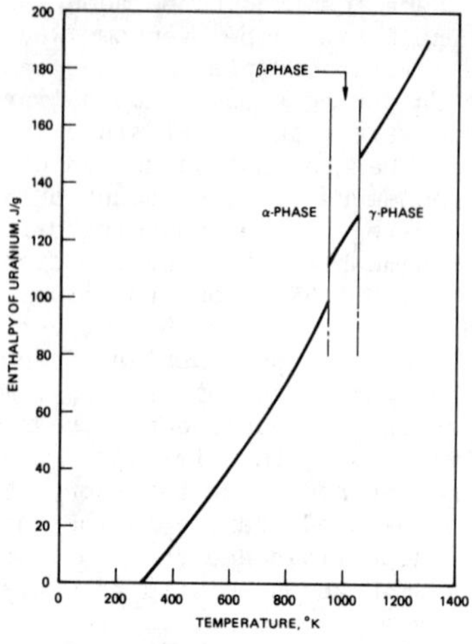

Figure 7. The significant contribution of the heats of transformation to the overall enthalpy of unalloyed uranium is summarized in this plot where 30% of the total heat that must be removed upon quenching is shown to be due to the phase transformations.

Y-WS 83-1165

MODELING REQUIREMENTS
EXTENSIVE PROPERTY CHARACTERIZATION AS A FUNCTION OF TEMPERATURE AND MICROSTRUCTURE

TRANSIENT TEMPERATURE DISTRIBUTION:

- THERMAL DIFFUSIVITY, α
- SPECIFIC HEAT, C_p
- DENSITY, ρ
- HEATS OF TRANSFORMATION, ΔH_{tr}
- TRANSFORMATION TEMPERATURE (CCT BEHAVIOR)
- SURFACE HEAT TRANSFER COEFFICIENT

RESIDUAL STRESS PROFILE:

- THERMAL EXPANSION COEFFICIENT, α
- YOUNG'S MODULUS, E
- POISSON'S RATIO, ν
- STRAIN HARDENING MODULUS, E_T
- YIELD STRAIN, ϵ_S
- ULTIMATE STRAIN, ϵ_{ULT}
- YIELD STRESS, σ_Y

Figure 8. A summary of the material properties that had to be measured as a function of temperature and microstructure for the uranium-0.75% titanium alloy for incorporation into the thermal and stress modeling computer programs.

that had to be determined for the model to accurately simulate the material behavior upon quenching. This property characterization has been completed. In the interim time, the model was tested and demonstrated for the unalloyed uranium case since data existed in the literature for the required properties. The necessity for having accurate data for the various properties can be demonstrated using Figure 9 which exhibits the yield strength for the U-0.75Ti alloy as a function of temperature and microstructure. The temperature regime of interest is in the vicinity of 615°C where the martensite transformation initiates upon cooling. If you extrapolate the yield strength curve for the parent gamma phase down to 615° and extrapolate the martensite (alpha prime) curve up to 615°, you will discover that the martensite has a yield strength almost 25 times stronger than the parent gamma. Coupling this information with the knowledge that a volume contraction of between 0.5% and 1.0% is associated with martensite formation, you would conclude that any type of asymmetric quenching condition would result in distortion of a quenching specimen and yield asymmetric residual stresses. Indeed, U-0.75Ti alloy specimens have been observed to bend toward the direction of maximum heat transfer. The analytic model will be shown subsequently to predict this behavior.

Parametric Studies

The purpose of the parametric studies was to demonstrate the capability of the Quench Simulator and to determine the most significant input data as evidenced by significant changes in the thermal or stress predictions. The operating parameters as well as material and boundary condition characteristics covered in this study include the effects of:

- -temperature-dependent properties,
- -heat of transformation,
- -water temperature,
- -jet impingement,
- -oxide coating,
- -immersion rate
- -boiling curve variations,
- -insulating the bottom.

Four separate geometric models were utilized in these studies depending on the parameters being studied. They were:

1. a one-dimensional radial (R) model,
2. a two-dimensional model in the radial (R) and axial (Z) directions,
3. a two-dimensional model in the radial (R) and angular (theta) directions,
4. a three dimensional model in the R, theta, and Z directions.

The cases that were simulated are summarized in Table I. The initial case served as the reference or base case to which all subsequent runs were compared. Not all these cases will be presented here but rather just a few representative ones to demonstrate the flexibility of the model. The detailed specifics of all the runs can be obtained in the reference by

Llewellyn, et al.[4] The specific geometry being simulated is a circular right cylinder as shown in Figure 10.

Figure 11 clearly shows the effect of running the thermal calculations with and without accounting for the heat of transformation in a one-dimensional analysis. The surface and centerline temperature profiles are significantly altered by the heat of transformation. Internal heat generation by phase transformation is definitively shown to be an important consideration when modeling phase decomposition especially in quench rate sensitive alloys like U-0.75Ti.

The evolution of the temperature and microstructure distributions in U-0.75Ti is depicted in Figures 12 and 13 respectively for a controlled immersion rate (6.35mm/s) quench process. The bottom surface is the one which makes contact with the quench medium first and the plots are for only half of the sample since the results are symmetric about the centerline. The isotherms shown in Figure 12 are of particular interest since they show a large thermal gradient evolving just below the upper end (refer to 60 sec plot) of the cylinder during the final stages of quenching. Due to the significant temperature dependence of the mechanical properties of this alloy plus the volume contraction associated with the martensite transformation, this region would be suspected to develop large residual stresses. Under fast immersion conditions, the U-0.75Ti alloy has been observed experimentally to develop centerline porosity and cracking in these regions exactly. Since all of the mechanical properties were not available until recently for the U-0.75Ti alloy, a computer simulation for the alloy could not be performed. However, the simulation of the quenching of unalloyed uranium under these conditions was possible since all of the necessary data was available. The allotropic phase transformations in unalloyed uranium also involve contraction and so provide a realistic comparison with the U-0.75Ti alloy behavior. The residual stress predictions for the radial and axial normal stresses are shown in Figures 14 and 15 respectively. These plots show significant residual tensile stresses (far in excess of the room temperature yield strength of unalloyed uranium) in the vicinity of the large thermal gradient area depicted in Figure 12. These high stress regions conceivably could initiate the centerline porosity and cracking observed in the U-0.75Ti alloy. These qualitative correlations between the analytic predictions of the Quench Simulator and actual experimental observations demonstrate the successfulness of the modeling endeavor. More quantitative correlations are anticipated when the actual surface heat transfer and mechanical property data for the alloy are incorporated into the Quench Simulator.

The influence of a 200 micron thick Zirconia coating is shown in Figures 16 and 17 where half of the specimen surface area was covered by the oxide. The isotherms (Figure 16) that evolve during quenching are asymmetric due to the effect of the zirconia layer on the surface heat transfer coefficient on the cylinders surface. The influence of these conditions on the nonuniform martensite evolution is seen in Figure 17. Similar effects are observed for any condition that causes an asymmetry to be developed in the surface heat transfer, e.g., a water jet spraying on one side of the sample only. The net effect on the residual stresses is similar in that isostress lines are skewed to one side of the sample resulting in distortion

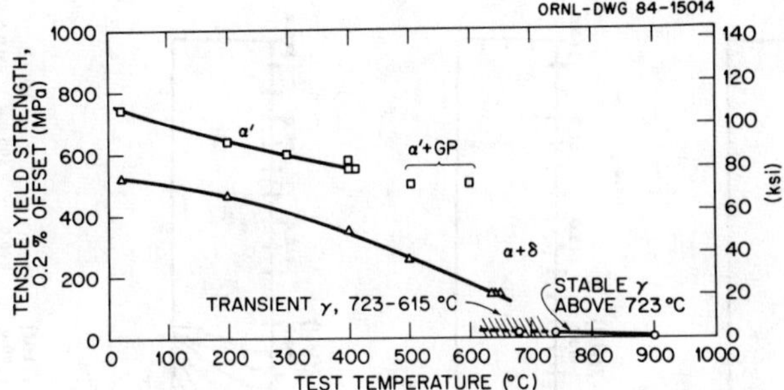

Figure 9. The yield strength of the uranium-0.75% titanium alloy as a function of temperature and microstructure is summarized in this plot.

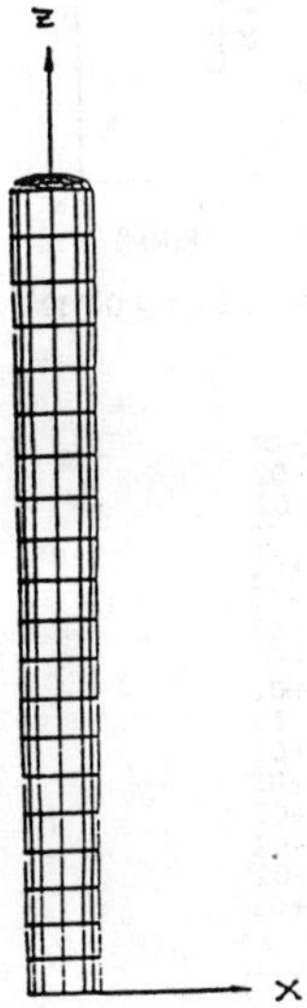

Figure 10. A schematic of the nodal mesh network for the cylindrical shape modeled in the parametric studies utilizing the Quench Simulator.

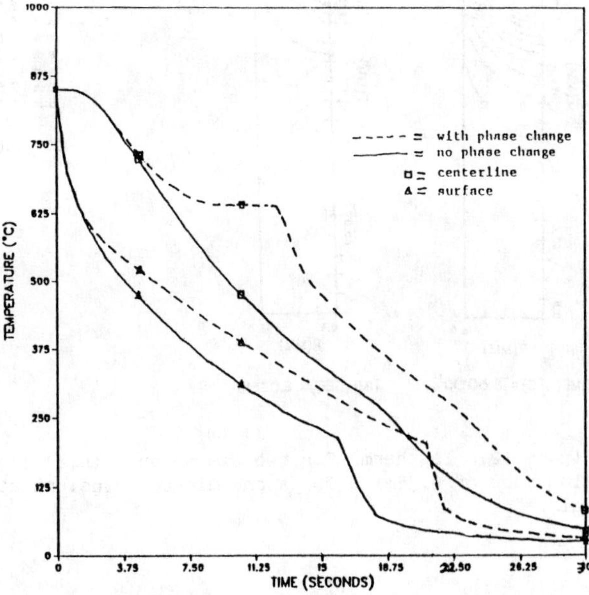

Figure 11. Comparisons of the surface and counterline temperatures as functions of time for a one-dimensional model with and without the heats of phase transformation.

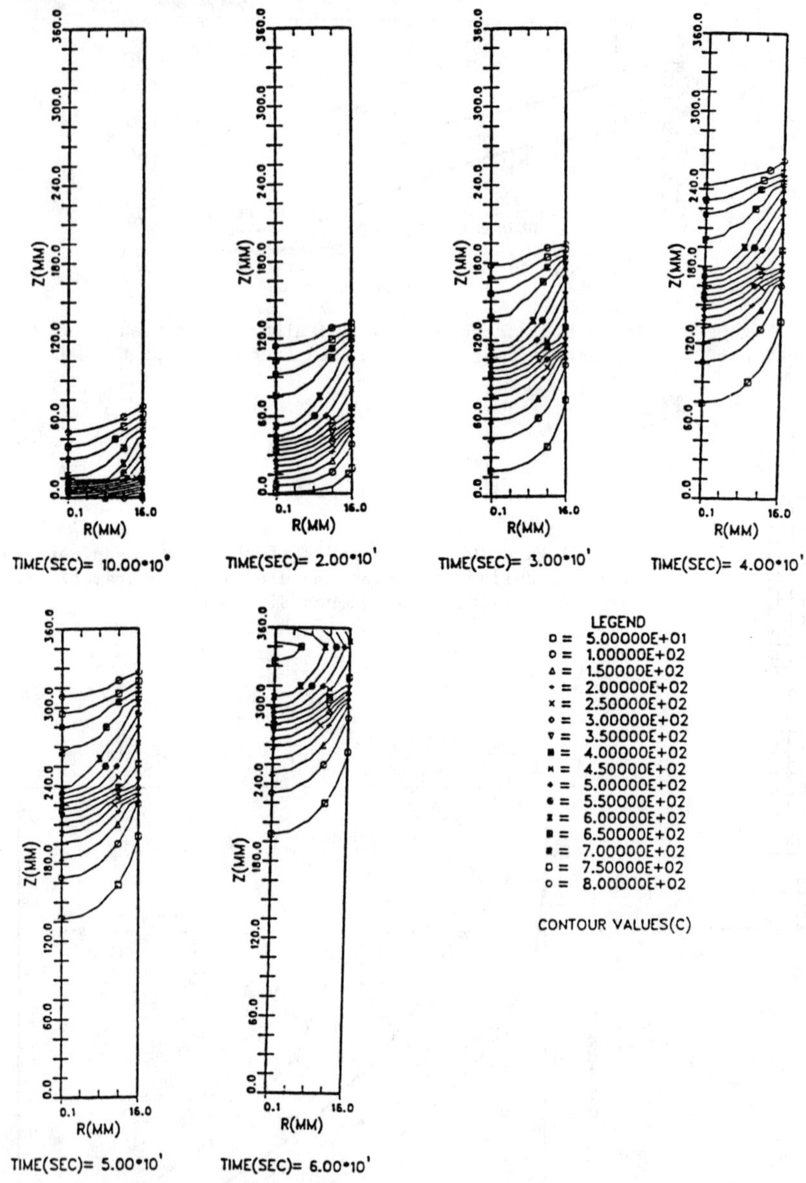

Figure 12. Isotherms for two-dimensional (R,Z) immersion model at an immersion rate of 6.35mm/s for a one minute transient at 10-second intervals.

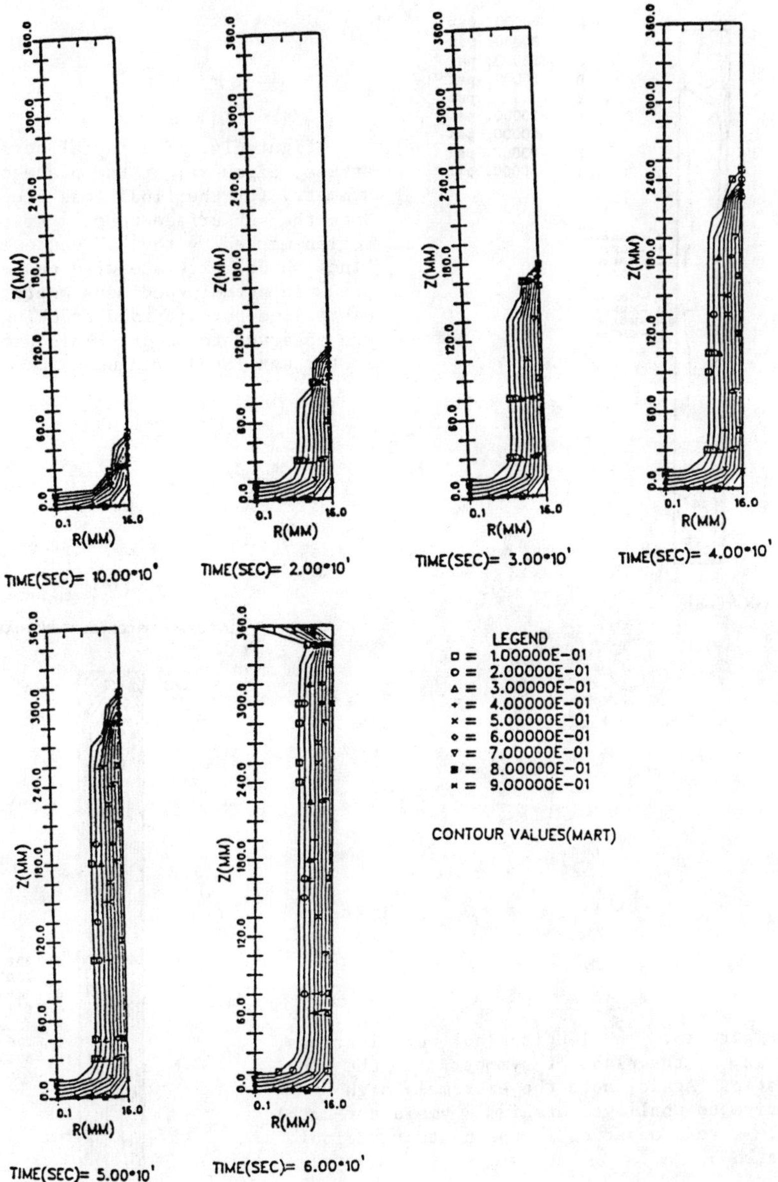

Figure 13. Martensite phase distribution for two-dimensional (R,Z) immersion model at an immersion rate of 6.35mm/s for a one minute transient at ten second intervals.

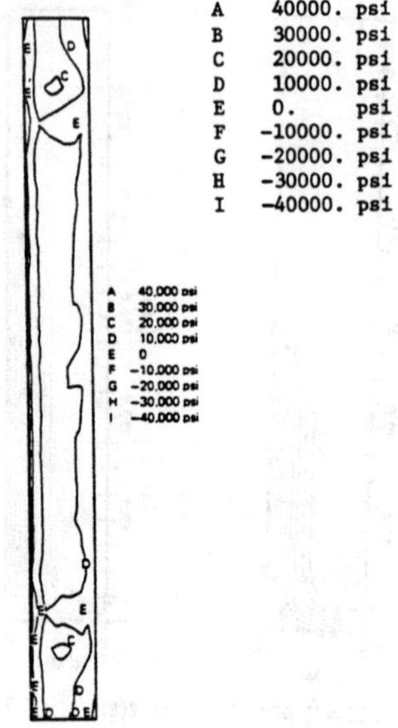

STRESS X-X AT LOAD 13

Figure 14. The radial normal stress, Sigma xx, at the plane of symmetry for the final load step. Note the subsurface tensile stress region marked by the "C" contour lines which correlate with those areas in actual specimens where centerline porosity and cracking are observed to occur. This plot is for unalloyed uranium.

ORNL–DWG 85-14649

PLOT OF STRESS CONTOUR PENETRATOR

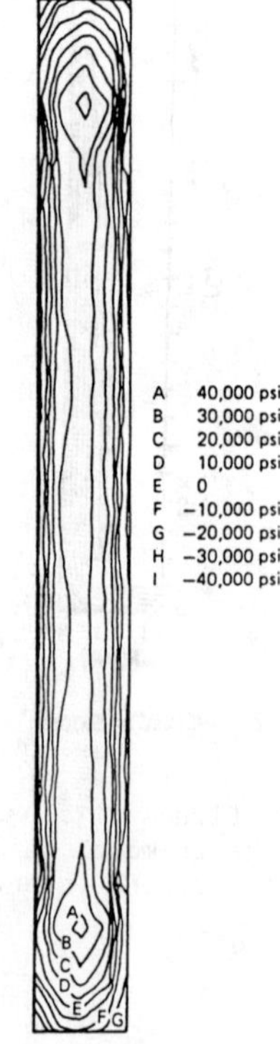

STRESS Z-Z AT LOAD 13

Figure 15. The longitudinal normal stress, Sigma zz, at the plane of symmetry for the final load step. Again, note the extremely high (relative to unalloyed uranium's yield strength) tensile stress observed in the contour region indicated by "A".

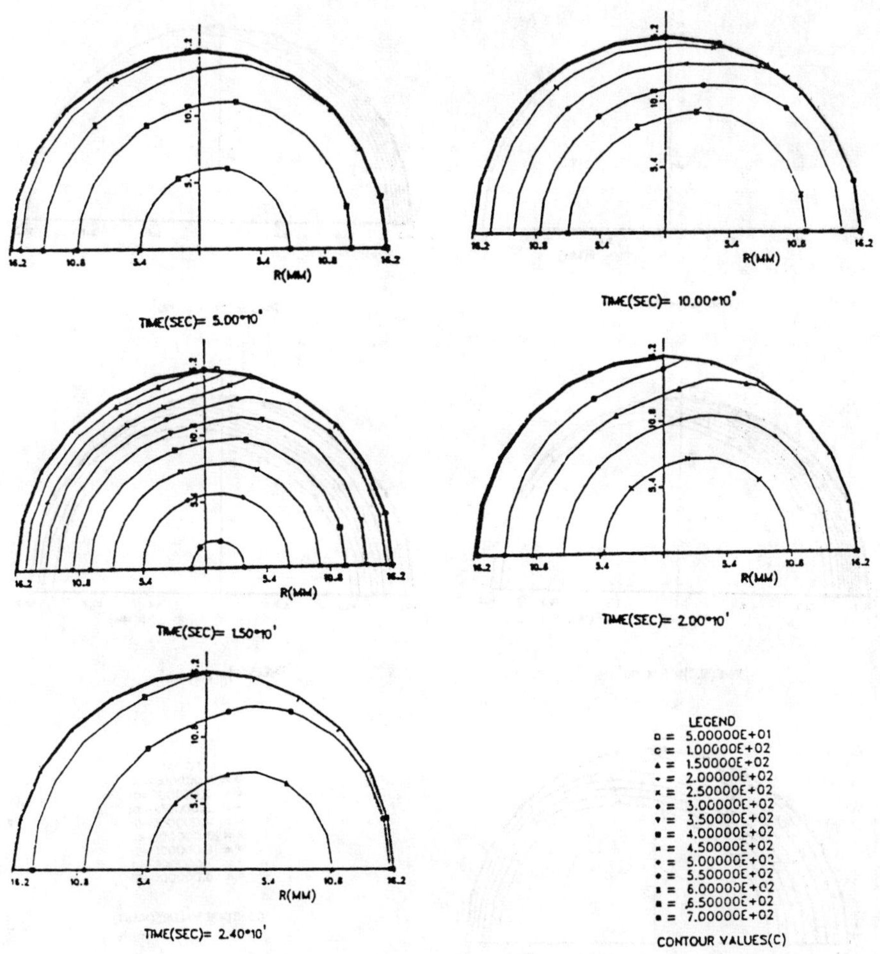

Figure 16. Isotherms for the two-dimensional (R,theta) model with standard boundary heat transfer coefficient and a 200 micron Zirconia coating over one half of the surface area (right half).

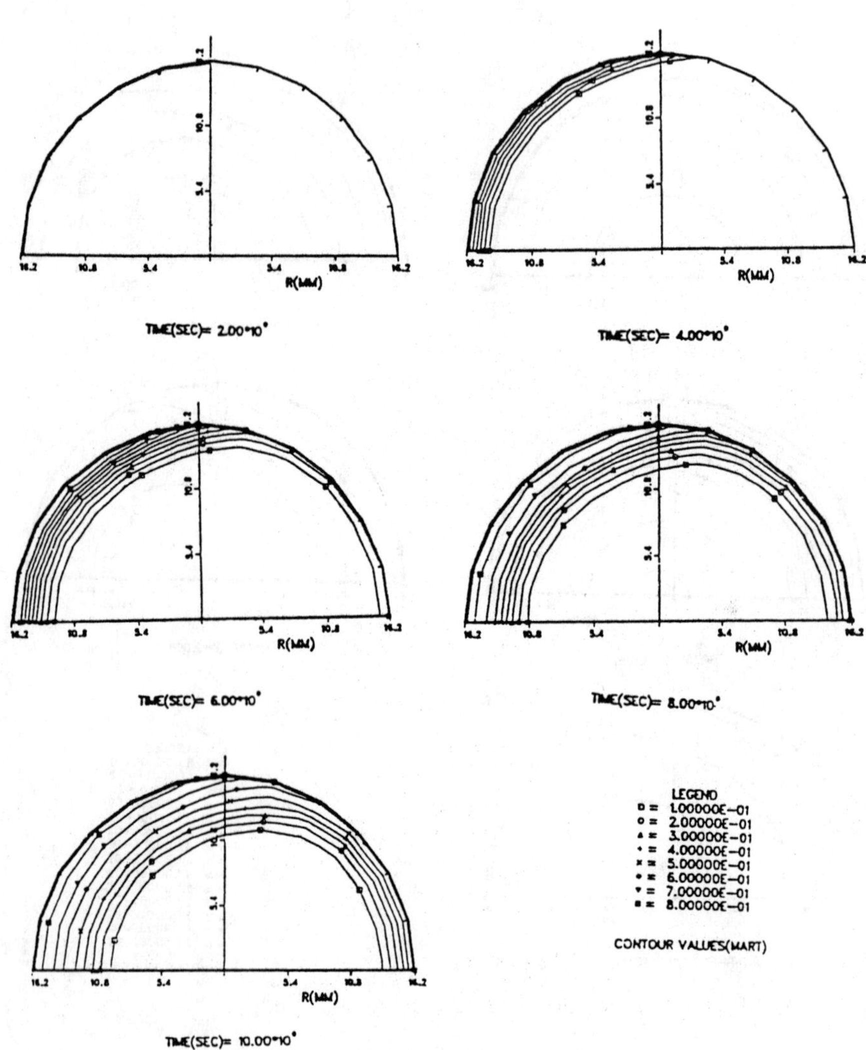

Figure 17. Martensite phase distribution for the two-dimensional (R,theta) model with standard boundary heat transfer coefficient and a 200 micron Zirconia coating over one half of the surface area (right half).

of the cylinder as shown in Figure 18. The distortion is manifested as a bow with the concave side being toward the region of maximum heat transfer for the uranium systems being studied. The distortions shown in Figure 18 have been amplified by a factor of ten for aid in visual presentation.

The final point of discussion is the relative importance of the magnitude of the heats of transformation for the phase transformations and of the surface heat transfer coefficients. Parametric studies were run for each of these variables to determine which parameter was more important to have absolute values for. The range of values covered in this study are listed in Table II. Figure 19 shows the results of this analysis. The variation of these two variables on the amount of martensite evolution was the criteria used for comparison of their effects. A comparison of the heat of transformation variation plots with the surface heat transfer variation curves of Figure 19 demonstrates that the surface heat transfer coefficient is the more important parameter. A factor of two change in the heat of transformation for the martensite reaction has only a minor effect on the amount of martensite evolution. On the other hand, a similar change in the surface heat transfer coefficient causes a change from half of the cylinder having essentially no martensite to a condition where the entire cylinder has at least twenty per cent martensite. The implications of both of these results are significant. For example, the martensite transformation in the U-0.75Ti alloy occurs extremely rapidly. Quench rates of 200°C per second are needed to obtain 100% martensite. Conventional techniques for measuring equilibrium phase change heats of transformation are not amenable to directly measuring this data for rapid transformations. Therefore, alternate indirect methods must be used to obtain this information. For this case, equilibrium thermodynamics was used using the principle that enthalpy is a state function and depends only on the beginning and final points. Therefore, knowing equilibrium phase change data and measuring the heat evolved during phase decomposition of the martensite to the equilibrium alpha plus delta, the heat of transformation for the martensite can be back calculated out. However, this technique may result in error for this data. This parametric study showed that the modeling of martensite evolution is not that sensitive to the magnitude of the heat of transformation and so error can be tolerated for this input data. On the otherhand, the surface heat transfer coefficient data has a significant effect and warrants experimental measurement of this information for the type of quench process being considered. This fact is important to realize since factors like agitation of the water quench bath or the use of a spray jet can vary the effective surface heat transfer coefficient by an order of magnitude.

Summary

A Quench Simulator has been developed which predicts the behavior of a metal during quenching. The actual non-linear, temperature- and microstructure-dependent physical, thermophysical, and mechanical properties were incorporated as input into the computer model as well as the continuous cooling transformation behavior and heats of transformation of the alloy. The results of several parametric studies have been presented which have shown the significant effect of asymmetric boundary conditions, and variations in the heats of transformation and surface heat transfer coefficient on distortion and martensite evolution. For unalloyed uranium and the U-0.75Ti alloy, the asymmetric boundary conditions were demonstrated

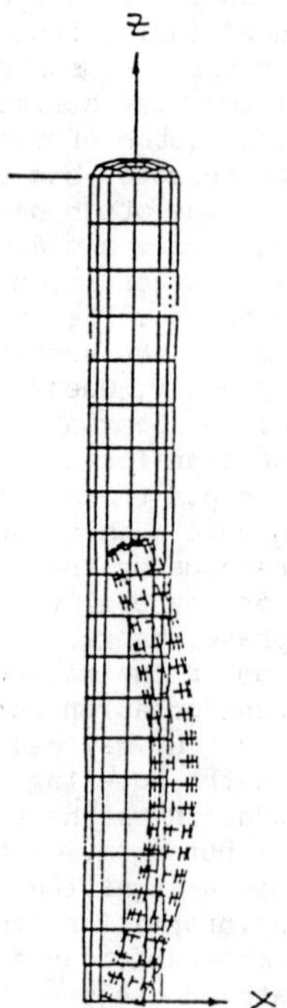

Figure 18. Final and initial configuration of the cylinder model for an asymmetric surface heat transfer condition. The region of maximum heat transfer is on the left. The deformed shape deformation strains (both thermal contraction and stress induced) have been magnified 10 times for ease of observation.

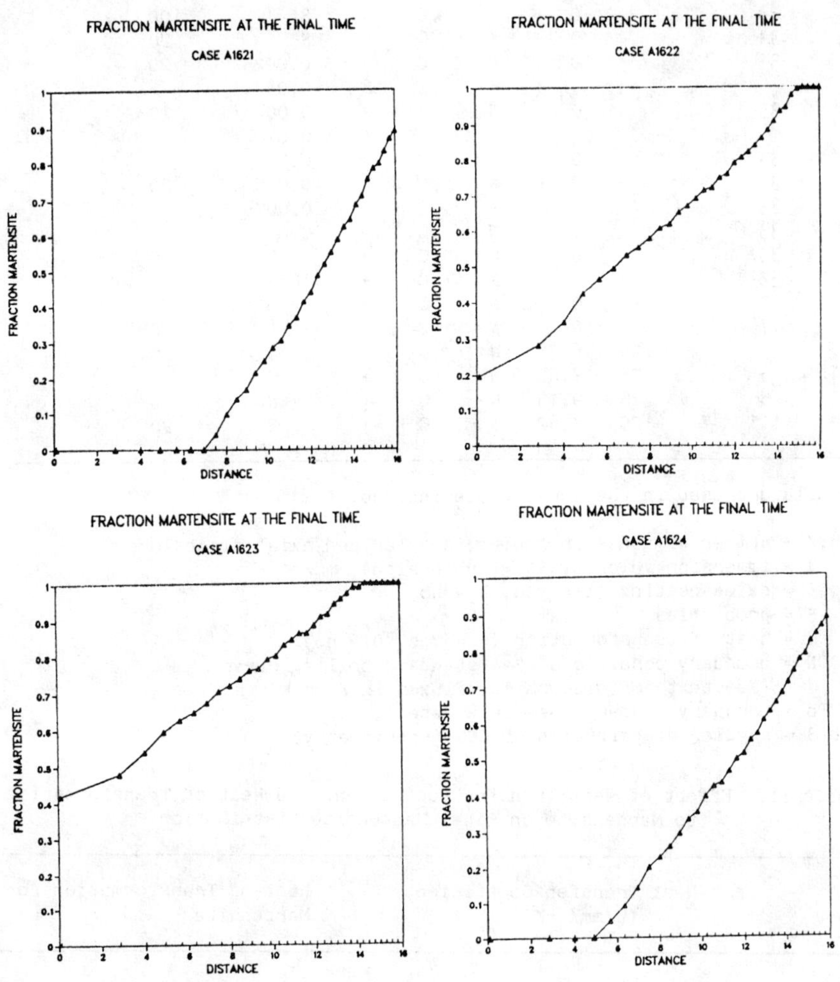

Figure 19. A parametric study showing the effects of varying the surface heat transfer coefficient and the heats of phase transformation on martensite formation for the U-0.75Ti alloy. A summary of the values for each case are presented in Table II.

Table I

Case	R	Th	Z	V	OC	P	H	BCN	Tb	Dis
base	33			0	N	C	+	S	20	
1	33			0	N	T	+	S	20	
2	33			0	N	C	0	S	20	
3	33			0	N	C	+	S	100	
4	34			0	Y	C	+	S	20	
5	33			0	N	C	+	0.0025/ 0.04	20	
6	33			0	N	C	+	0.0025/ 0.01	20	
7	33			0	N	C	+	0.0025	20	
8	33			0	N	C	+	0.005	20	
9	33			0	N	C	+	0.0075	20	
10	33			0	N	C	+	0.02	20	
11	33			0	n	C	+	5 S	20	
12	33			0	N	C	+	10 S	20	
13	17	13		0	N	C	+	S	20	
14	17	13		0	N	C	+	S	20	
15	5		25	6.35	N	C	+	S	20	
16	5		25	7.62	N	C	+	S	20	
17	5		25	9.14	N	C	+	S	20	
18	11	7	25	7.62	N	C	+	S	20	

Nomenclature used in the above table include.

R,Th,Z = number of nodes in radial, angular and axial directions
V = immersion velocity (0 = submersion), mm/s
OC = oxide coating (Y = yes, N = no)
P = properties
H = heat of transformation (+ = yes, 0 = no)
BCN = boundary conditions (S = standard boiling curve) (see text for discussion, fluxes in W/mm.K)
Tb = boundary temperature (bulk water), C.
DIS = angular distribution of BCN (nonsymmetry)

Table II. Effect of Heat Transfer Coefficient and Heat of Transformation to Martensite on Final Martensite Distribution

Case	Heat Transfer Coefficient (W/mm^2 -K)	Heat of Transformation for Martensite
A1621	0.0025	0.5409
A1622	0.0050	0.5409
A1623	0.0075	0.5409
A1624	0.0025	0.2705

Case A1621 is approximately the same as the base case A1601.

to cause distortion of the cylinder with bowing occurring toward the maximum heat flux region. Similarly, the heat transfer coefficient was shown to have a much more significant influence on the predictions for martensite formation than the heat of transformation for the martensite reaction. Experimental measurement of the heat transfer coefficient for an immersion quench has been completed and will be incorporated into the computer model in the near future to replace the steady state heat transfer data obtained from the literature.

This quench code has the flexibility to model quenching of other metallic systems since the material properties and CCT behavior are input as subroutines to the main code. This type of analysis will be extremely beneficial as part of a CAD/CAM approach to part and manufacturing equipment design.

REFERENCES

1. D. C. Elrod, G. E. Giles, and W. D. Turner (1981), "HEATING6: A Multidimensional Heat Conduction Analysis With the Finite-Difference Formulation," Union Carbide Corporation, Nuclear Division, Oak Ridge, Tennessee, October, ORNL/NUREG/CSD-2/V2, Section F10.

2. "ADINAT - A Finite Element Program for Automatic Dynamic Incremental Nonlinear Analysis of Temperatures," ADINA Engineering AB Munkgaten 20D S-722 12 Vasteras, Sweeden.

3. J. V. Beck, "Surface Heat Flux Determination Using an Integral Method, "Nuclear Eng. and Des., Vol. 7, 1968, pp. 170-178.

4. G. H. Llewellyn, G. A. Aramayo, K. W. Childs, G. M. Ludtka, and M. Siman-Tov, "Computer Simulation of Immersion Quenching of Uranium-0.75% Titanium Alloy Cylinder," Martin Marietta Energy Systems, Inc., Report Y-2355, in print (1986).

Subject Index

Author Index